THE CHINESE SUGAR-CANE.

THE

CHINESE SUGAR-CANE;

ITS

History, Mode of Culture,

MANUFACTURE OF THE SUGAR, ETC.

WITH

REPORTS OF ITS SUCCESS IN DIFFERENT PORTIONS OF THE UNITED STATES, AND LETTERS FROM DISTINGUISHED MEN.

WRITTEN AND COMPILED BY

JAMES F. C. HYDE,
OF WALNUT GROVE NURSERY, NEWTON CENTRE, MASS.

BOSTON:
PUBLISHED BY JOHN P. JEWETT & COMPANY.
CLEVELAND, OHIO:
HENRY P. B. JEWETT.
1857.

STEREOTYPED BY
HOBART & ROBBINS,
New England Type and Stereotype Foundery,
BOSTON.

PREFACE.

Few subjects are of greater importance to us, as a people, than the producing of sugar; for no country in the world consumes so much as the United States, in proportion to its population. It is a subject of special interest at this time, on account of the great advance that has taken place in the price of this very important product.

We have been hoping for years to obtain a plant which would produce sugar in the northern portion of our country; and it is supposed by many that we have at last succeeded. But whether we have or not, the subject is exciting a great deal of interest, especially with the writer; and his attention has been given to the new plant,—not, however, without fear that it would prove, like many other new things, comparatively worthless. Having ascertained certain facts in regard to it, he was inauced to give them to the public through the newspapers, supposing that that would be the last of it, so far as he was concerned. But, to his surprise, letters began to pour in, at the rate of three or four a day, from all parts of the country, from Maine to Minnesota, asking for further information, and for seed of the plant. These letters were answered, and seed sent free of charge, until they came so thick and fast, he was obliged to say that he could not answer them in detail.

Finding there was such a desire to obtain information, on the part of the public, he was induced, at the suggestion of a friend, to get up this little, unpretending volume. No merit is claimed for it, other than that it is the truth, and the whole truth, so far as the experience of the writer goes, and so far as he has been able to obtain information from other sources; for he has carefully avoided everything that did not seem to be well authenticated.

The writer hopes and believes this little work will prove useful to those who wish for information in regard to the new plant of which it treats. He has given all the information that could be obtained on the subject. The work was attended with some difficulties, owing to the fact of the recent introduction of the plant, and consequently the short time there has been to try experiments with it. The writer feels a deep interest in this subject, and that has led him to bring this before the public. But, while he gives the result of his own experience, he also gives a statement of most of the experiments that have been made in the country. For an account of these he is indebted to Richard Peters, Esq., who furnished a detailed report of his trial of the cane; D. Redmond, editor of the *Southern Cultivator;* the Patent Office Reports, and some of the agricultural papers North and South.

The object of this work is to supply the public with accurate knowledge concerning this new and valuable plant, — Chinese Sugar-Cane. How far he has accomplished that object the reader must judge.

J. F. C. H.

NEWTON CENTRE, *Dec.* 20*th*, 1856

THE

CHINESE SUGAR-CANE.

The great value and extensive consumption of the products of the sugar-cane lead us to feel a deep interest in its cultivation, and especially now that prices are so high; and while we believe that we have, in the new Chinese sugar-cane, a plant adapted even to the most northern of the United States, and one too that can be grown so easily, and yield so richly. Sugar is no longer a mere luxury, denied to all but the rich and great, as it was once, but is used by all classes. It may not be uninteresting to the reader to give something of the history and origin of the sugar-cane. All the evidence goes to show that China was the first country that cultivated it, and manufactured sugar; and not only were the Chinese the first, but there is good reason to believe that they enjoyed its use many centuries before it was generally known and used in Europe. Indeed, it would further seem that they not only possessed the art of extracting the juice, but a knowledge of the whole process, down to refining sugar. Strange as it may seem, it was a long time in finding its way over the different countries where it is now so profitably cultivated. When first known, it went by the name of Indian salt, and under that name it was sent abroad from China to India and Arabia, and thence to Rome and Greece, among the costly spices, and was considered a rare luxury. The cultivation of the plant gradually extended over the different countries of Europe.

It is supposed that it was known in the south of Europe as early as the ninth century, for there is evidence that it was cultivated at Sicily and the islands in its vicinity; but it was not until the thirteenth century that the cane became generally known and cultivated on that continent. It has finally extended over most of the civilized world where the climate is adapted to its growth. For some time after the introduction of sugar into Europe it was used only on great occasions, such as feasts, and for medicines; and in a different form from what it is now commonly used,—more like our candy;—and it is within a hundred years that the sugar-cane was first brought to this country and cultivated to some extent in Louisiana, on the very place where the city of New Orleans now stands. But the planter was not able to do any more than make syrup, not perhaps ever expecting to make sugar; yet we see since that time hundreds of thousands of hogsheads have been produced every year. Why may not the same results follow the introduction of the new Chinese sugar-cane? The following account is given of the first experiment of sugar-making in that state:

"Towards the close of the last century," says the highest authority, "a gentleman residing in the vicinity of New Orleans determined to attempt the manufacture of sugar. The crop was properly increased, the machinery procured, and a sugar-maker procured from the West Indies. The result of the experiment was anxiously looked for by the whole surrounding country. The inhabitants of New Orleans and its neighborhood assembled in great numbers, but remained outside of the building, probably through fear that the experiment would not succeed. The *strike* was made amidst profound silence; when the 'second' was thrown into the coolers, the sugar-maker announced to the anxious crowd, in technical language, 'It grains!' shouts of joy rent the air, and the news spread with rapidity that the juice of the cane grown in lower Louisiana had been manufactured into crystallized sugar, and a new impulse was given to the cultivation of cane." Year by year this crop has increased in value, and has now become a very valuable one, and a great article of commerce. What variety of cane was first brought from China and India, I do not know, but probably one of the sorts that are now cultivated in the West Indies, and in our own sugar-producing state, Louisiana.

It may be well to give a few statistics concerning the amount of sugar consumed in this country, and other facts connected with the trade in this article. There is, annually, consumed in the country, upwards of 800,000,000 pounds, or about 30 pounds to each person; which is certainly a large amount, much larger than is consumed in any other country, in proportion to its inhabitants. In Great Britain, each person consumes 24 pounds; Belgium, 18; Holland, 17; France, 8; Denmark, 6; Sweden, 4; Russia, 2½. This shows conclusively that the people of the United States consume more sugar than those of any country. It is considered indispensable by every one, and its use judged conducive to health. According to the Patent Office report of 1853–4, by the census returns of 1840, the amount of cane-sugar made in the Union, that year, was 119,995,104 pounds; in 1850, 247,577,000 pounds, showing an increase of 127,581,896 pounds, besides 12,700,896 gallons of molasses. The amount of cane and maple sugar made in the United States in 1853–4 may be estimated at 545,000,000 pounds, which at six cents — and it is worth much more than that — would be worth $32,700,000, besides 14,000,000 gallons of molasses and syrup, which, at 30 cents, would be worth $4,200,000. A large part of the sugar which is produced in this country comes from Louisiana, as will be seen by the following returns for the years 1853–4, — one year's product, — in all, 449,324 hogsheads, or about 495,156,000 pounds. Thus, it will be seen that Louisiana supplied much more than was produced in the whole country in 1850. But we see, by late returns, that the crop has fallen off, so that in 1855 it was only 231,427 hogsheads, or but little more than half what it was in 1853; and this probably is one great reason why sugar is so much dearer than formerly.

According to the statement of the New York Shipping and Commercial List, the total importation of foreign sugars into the United States for the year ending December 31, 1855, was 382,786 hogsheads, of 1200 pounds each; in 1854, 309,726 hogsheads. According to Hunt's Magazine, the total decrease of cane-sugar in 1855, as compared with 1854, was $1\frac{5}{8}$ per cent. The amount of sugar consumed in the United States in 1855 was about 766,000 hogsheads, of 1200 pounds each. The greatest amount produced in Louisiana in one season being 449,324 hogsheads, of 1000 pounds each. In addition to this, the amount of maple-sugar made in the United States in 1850 was 34,253,436 pounds. Of this, New York produced 10,357,484; Vermont, 5,980,955; Ohio, 4,588,209; and so on, every state producing some.

I might go more extensively into statistics, if necessary, to show the importance and value of this crop; but deem it unnecessary, as every person is compelled to own that it is second to few other crops. And though we see, by the facts presented, that the culture of sugar-cane and the manufacture of sugar from the maple has greatly increased within the past few years, and that it must naturally increase in time to come, yet the demand has been greater than the supply, or, in other words, the demand being so active and the supply limited, prices have greatly advanced, so that sugar that was sold ten years ago for six cents is now worth nine and ten cents, or nearly double what it was then. There are several reasons for this: much more is used than formerly for preserving fruits, for confectionery, &c.; less sugar is produced in the British West Indies than formerly; new markets have been opened and have been supplied; while hard winters in Louisiana have, in some instances, killed out the cane, and there have been many other things to contend with in its growth, all of which have tended to diminish the supply of this valuable article. There are other well-known facts connected with sugar-growing in Louisiana, which show the difficulties and obstacles there are in the way of sugar-growing in that state. One is, the great expense they must be at in draining the land, and preparing it for the growth of the cane. This objection will hold good of much of the land on which sugar-cane is grown; then, when the cane is matured, they must reserve about a fourth of their entire crop for next year's setting. Their cane is grown from cuttings or joints of last year's growth, unlike the West India cane, which lives in the ground year after year, or the Chinese sugar-cane, which produces seed from which it may be grown ; — the cane hitherto cultivated never seeds in this country, and rarely anywhere. It has also deteriorated from being reproduced in this way year after year, from cuttings, so that it takes more acres to yield the same amount of sugar than formerly. Sugar-cane will not flourish on a wet soil. In regard to maple-sugar, it may be said that the supply will probably be limited ; and even allow that it should continue for many years to come as it is, or even increase, how much would it do towards supplying a constantly increasing demand? The same argument will apply, and perhaps with greater force, to the manufacture of beet-sugar, to which considerable attention has been paid in France, though but little has been done in our own country. In 1810, when Napoleon the Great did everything in his power to encourage the cultivation of the sugar-beet, for the manufacture of sugar, there was produced that year 2,000,000 pounds, or about one fifty-eighth part as much as France consumed. Subsequent to that, its manufacture increased to an extraordinary extent, and annually yielded 24,000,000 pounds. When a tax was laid upon domestic sugars, it again decreased. But, for the year preceding the first of September, 1853, there was manufactured in France, from the sugar-beet, 165,680,790 pounds. A great portion of the beet crop of France is now used for the manufacture of brandy, the grape crop having partially failed. The reputation and demand of French brandy is such, that it becomes profitable to use the beet crop for this purpose, rather than to make sugar. By this France becomes dependent on foreign countries for a large part of the sugar which it consumes. The beet culture for sugar has been extended over Germany, Belgium, and other European states.

A writer in the *Boston Herald*, speaking of the deficiencies of the sugar crop of our own and other countries, and showing that the crop of Louisiana will be very much smaller than last year, goes into some figures to show this fact, and quotes from the *Philadelphia Journal*, which says: "In all probability, before next summer, the sugar sold at eight cents a pound in 1854 will reach at least double that sum. John Brown, the laborer, must therefore prepare to sweeten his tea with steam-syrup molasses."

He also speaks of the comparatively small yield, and the entire inadequacy of the maple-sugar to meet this demand, and closes by recommending the culture of the sugar-beet. He cites the success of the French in this branch of sugar manufacture as an example for us, and thinks the agricultural department of the Patent Office ought to take some steps to bring about "a consummation so devoutly to be wished,"— a decrease in the price of sugar and molasses. He further says: "The annual yield of beet-root sugar in France averaged for a number of years upwards of 150,000,000 pounds. This immense production, at thirteen cents a pound, the price we now pay for a decent article of Havana sugar, at retail, gives the handsome sum of about $20,000,000 added yearly to French industry."

It may not be out of place here to say that the sugar-beet culture has been attempted, and thus far failed; and I very much doubt if it can be revived and be made profitable to compete with the new Chinese sugar-cane, which is now being brought forward to supply, if possible, the deficiency in the sugar and syrup crop of our country. I have no doubt, if half the time and money are spent in perfecting the manufacture of sugar from the new cane that have been spent in France on the sugar-beet culture, and manufacture of sugar therefrom, surprising and very gratifying results will follow. For, not only will it, in my opinion, take the place of the old varieties of cane in Louisiana, and thus render the crop of that state much larger and more sure, but it will be grown in almost every state and territory in our widely-extended country, either for the production of sugar or syrup, for both of which I shall attempt to show it is adapted. Now, what do these facts concerning the sugar crop show? Do they not show most conclusively that we must, if possible, increase in some way the annual production of sugar? — and the question is, how shall it be done? The only remedy is to find saccharine plants adapted to the temperate zone, so that they may be profitably employed in the production of sugar. If this can be done, and the farmers of each state raise and make their own sugar and molasses with the same ease with which they grow wheat and other grain, and manufacture it into flour, then these articles will be so extensively raised that the supply will be adequate to the demand, and prices will be reduced as they should be, so that all, both poor as well as rich, may enjoy the benefit of them.

Most of us have hitherto despaired of finding such a sugar-yielding plant that could be grown in a northern climate. Perhaps it is not yet found; but we may hope, and not without strong reasons, that the Chinese sugar-cane is just the article calculated to supply the want. So that we think the day is not far distant when sugar and syrup or molasses enough will be produced, even in the New England States, to supply our wants, and thus relieve our country of the heavy tax it has paid foreign countries for this very useful and necessary article.

Having shown to some extent the value of the sugar crop, the increasing demand for sugars and molasses, the reasons why we cannot depend upon the present sugar-growing countries for a supply at fair prices, the difficulties they have to overcome in its cultivation, and, lastly, the necessity there seems to be of procuring some new plant that will grow everywhere, and produce the rich saccharine matter we so much desire, I am prepared to introduce to the attention of the agriculturists of our country the *new Chinese Sugar-Cane*, of which we shall give a short history, and then detail the results that have followed the experiments that have been made with it during the past two years.

D. J. Browne, Esq., who is connected with the Patent Office, and who, I believe, first introduced the Chinese sugar-cane into this country, gives us the following full and valuable account of it, in the report of 1854 He says it is "a new gramineous plant, which seems to be destined to take an important position among our economical products; was sent some four years since from the north of China, by M. de Montigny, to the Geographical Society of Paris. From a cursory examination of a small field of it growing at Verrieres, in France, in autumn last, I was led to infer that, from the peculiarity of the climate, and its resemblance in appearance and habit to Indian corn, it would flourish in any region wherever that plant would thrive. But how far it will subserve the purpose ascribed to it in France, should it even succeed in every part of the United States, can only be determined by extended experiments.

"There appears to be a doubt among the scientific cultivators in Europe as to the botanical name of this plant. *Holcus Saccharatus*, which is evidently an error, has been provisionally adopted by M. Louis Vilmorin, of Paris; but, as the term is already applied to our common broom-corn, if not to other species, this name cannot with propriety be retained. Mr. Leonard Wray, of London, who has devoted much time and attention to the cultivation of this plant with a view of extracting sugar from its juice, at Cape Natal and other places, informed me that in the south-east part of Caffraria there are at least fifteen varieties of it, some of them growing to a height of twelve or fifteen feet, with stems as thick as those of the sugar-cane. M. Vilmorin also says that in a collection of seeds sent to the museum at Paris, in 1840, by M. de Abadie, there were thirty kinds of sorghum, among the growth of which he particularly recognized several plants having stems of a saccharine flavor. Thus it will be seen that there is much cause of confusion, and a necessity for a critical examination of the subject. I would state, however, that Messrs. Vilmorin and Groenland are engaged conjointly in the cultivation, and in determining the properties of this and the allied species, and we have every reason to hope that their researches will enable us soon to know their botanical types.

* * * * * *

"Sorgho Sucre is a plant which on rich land grows to the height of from two to three or more yards. Its stems are straight and smooth, having leaves somewhat flexous and falling over, greatly resembling Indian corn in appearance, but is more elegant in form. It is generally cultivated in hills containing eight or ten stalks each, which bear at their tops a conical panicle of dense flowers, green at first, but changing into violet shades, and, finally, into dark purple at maturity. In France it is an annual, where its cultivation and period of growth correspond to those of Indian corn; but, from observations made by M. Vilmorin, it is conjectured that, from the vigor and fulness of the lower part of the stalks in autumn, by protecting them during the winter, they would produce new plants the following spring. If cultivated in our Southern States, it is probable that the roots would send forth new shoots in spring, without protection, in the same manner as its supposed congener, the Dourah corn. At the North, the maturity of the seed probably would be more certain if planted in some sheltered situation; but, if the object of cultivating be for the extracting of sugar, or for fodder for animals, an open culture would be sufficient, where the soil is rich and light, and somewhat warm. According to the experiments of M. Ponsart, the seeds vegetate better when but slightly covered with earth. M. Ledocte proposes to associate with the plant another of more rapid growth, such as lettuce, or rape, in order that the laborers may distinguish the young sorgho from grass, which it greatly resembles in the early stage of its growth. Any suckers, or superfluous shoots, which may spring up in the course of the season, should be removed.

"The great object sought in France, in the cultivation of this plant, is the juice contained in its stalks, which furnishes three important products: namely, sugar, which is identical with that of cane; alcohol, and a fermented drink analogous to cider. This juice, when obtained with care in small quantities, by depriving the stalk of its outer coating, or woody fibre and bark, is nearly colorless, and consists merely of sugar and water. Its density varies from 1.050 to 1.075, and the proportion of sugar contained in it from ten to sixteen per cent., a third part of which is sometimes uncrystallizable. To this quantity of uncrystallizable sugar this juice owes its facility of readily fermenting, and consequently the large amount of alcohol it produces, compared with the saccharine matter observed directly by the saccharometer. In so far as the manufacture of sugar is concerned, this plant appears to have but little chance of success in a northern climate, as a large proportion of that which is uncrystallizable is not only a loss in the manufacture, but an obstacle to the extraction of what is crystallizable.

"It must not be understood, however, that the produce of this plant is unprolific or difficult to obtain, but that, all things being equal, its nature renders it more abundant in alcohol than in sugar. Yet it would be very different in the warmer climate at the South, where the sugar-cane is difficult to be obtained, in requiring protection from frost. From experiments made by M. Vilmorin, on some dried stalks of sorgho sent from Algeria, it proved that the product of sugar obtained from them was infinitely superior to that produced from the same plant which had been cultivated near Paris. I was also informed by Mr. Wray, who experimented upon the juice at Natal, that the proportion of crystallizable sugar quite predominates where the climate allows the plant fully to mature. The chief advantage of the sorgho, as a sugar-plant, is the facility of its cultivation, and the easy treatment of the juice. It is thought that the rough product may surpass that of the sugar-cane in those countries where the latter is an annual, and, like which, its stalks and leaves will furnish an abundance of nutritious forage for sustaining and fattening animals. As the molasses, too, is identical with that manufactured from the cane, it may be used in the distillation of rum, alcohol, and a liquor called 'tafia,' which resembles brandy.

"The greatest difficulty to be apprehended, probably, would be the preservation of the stalks from fermenting, owing to the short time left to the manufacture. This, however, might be obviated, as Mr. Wray informed me that, in the neighborhood of Natal, the Zoulous-Caffers preserved it for a long time by burying the stalks in the ground, notwithstanding the climate of their country is very warm and damp. It will also be observed, that in the manufacture of brandy, or alcohol, the uncrystallizable sugar can be turned to account, which in a measure would otherwise be lost. Another advantage consists in the pureness of the juice, which, when thus converted, from the superiority of its quality, can immediately be brought into consumption and use. The alcohol produced by only one distillation is nearly destitute of foreign flavor, having an agreeable taste, somewhat resembling noyau, being much less ardent, or fiery, than rum.

"One of the points M. Vilmorin was desirous of establishing was, at what period of the growth the stalks began to contain sugar, and, consequently, when its manufacture should commence. He came to the conclusion that it coincided with the putting forth of the spikes; but the proportion of sugar in the stalks continued to increase, until the seeds were in a milky state. In the plant in flower, he observed that the amount of sugar diminished in the merithalles (parts of the stalks between the nodes, or joints), the nearer they were to the top; and also the lower part of each merithalle contained less saccharine matter than the upper. In consequence of this, and owing to the smallness and hardness of the lower knot, the centre of the stalk is the richest portion. He was inclined to the opinion that, at a later period, the merithalles lower down the stalk are impoverished in the amount, if not in the quality, of the sugar they contain. The ripeness of the seeds does not appear much to lessen the production of sugar, at least in the climate near Paris; but in other countries where it matures when the weather is still warm, the effect may be different. According to the report of M. de Beauregard, addressed to the 'Comice de Toulon,' the ripening of the sorgho in that latitude had no unfavorable effect; and he considers the seeds and the sugar as two products to be conjointly obtained. On the other hand, Mr. Wray says the Zoulous-Caffers are in the habit of pulling off the panicles of the plant the moment they appear, in order to augment the quantity of saccharine matter in the stalks. This question may be of some importance in our Southern States, should this plant supersede in any manner the sugar-cane. Having considered some of the probabilities of this product in an economical point of view, it remains only for me to recommend it to the attention of others who may have opportunities to cultivate it, and the means and talent to prove or refute, by direct experiments, its worth."

How far this new cane will stand the northern winters, yet remains to be proved. But it will be seen that it is not so important that this cane should stand the winters, for it can be grown readily from seed, without fear of deterioration, as in case of the old cane grown from cuttings, which produces less and less every year, showing most clearly that it cannot be depended upon for years to come. In fact, so apparent has this become, that fields that formerly produced 3000 to 4000 pounds of sugar to the acre now only produce, with equally good treatment, on an equally good soil, from 500 to 1000 pounds. And in order to preserve the sugar-fields from complete barrenness, the government has fitted out and placed at the service of the Patent Office a vessel which is to be sent out with a competent agent to procure sugar-cane cuttings from abroad, to stock anew the plantations of the South.

The new cane, as will be seen, is being fully tested in France by men competent to determine its value, and to whom the public will look with interest, while we shall watch with still greater interest the results of other experiments that will be made next season in our own country. It has been found, by careful experiment, to yield not only sugar and syrup, but alcohol; and the juice, when fermented, yields a drink much like cider; when set with alum, the juice of the husk is said to be good for dyeing, giving a permanent red; the trash, or waste, after it has been crushed and the juice expressed, will make a good article of paper, while the seed that the plant yields possesses fattening properties like rice, and can be profitably fed out to cattle, swine, &c. And this is not all; it will take the place of all other things for fodder for cattle, either to be fed green or dry, all of which properties will be treated more in detail hereafter.

First, in regard to obtaining sugar from this cane. But little has been done yet, it is true, though sugar has been made from it both in France and in this country, so that the matter is not at all in doubt; the only question being, whether it will supersede the old cane in the South, and can be profitably cultivated at the North. The writer made a small quantity of sugar this season, which, though of a dark color, for the want of knowledge as to the course that ought to be pursued, yet fully proved to his mind that all that is wanted is experience, to obtain sugar of the best quality, and in liberal quantities. It is thought by some that it can never be profitably raised for this purpose north of New York: time alone must determine this. If there were never a pound of sugar made north of that state, still the cane would be of immense importance to the North, on account of its other valuable properties. But it is presumption to say that sugar cannot be made from it; for, if such syrup can be produced as the writer will show he has obtained from this plant, then it must follow that sugar can also be produced. I am perfectly satisfied that it may be profitably grown for sugar making. In regard to the syrup, I can speak with great confidence from observation and experience, as well as from the results that have attended experiments in different parts of the Union. The juice yields from a fifth to a fourth of its bulk in good syrup; and such syrup as will make one wish at once for the griddle-cakes on which to test it. In proof of its quality, we give the following from the *Daily Evening Traveller:*

"MASSACHUSETTS MOLASSES.—We are indebted to J. F. C. Hyde, of Newton Centre, for a specimen of molasses which he has manufactured from the Chinese sugar-cane grown upon his farm in that town. It is equal to the best syrup; in color of a light brown, and of an excellent flavor."

If any further proof is necessary, I will give the words of an eminent merchant of Boston, who tested the syrup made by me,—a gentleman who is fully competent to judge, it having been a great part of his business to import and sell sugars and molasses. He said, after testing it, that it had "a peculiar fruity, cane flavor, and was a most splendid article," and wanted to know where I obtained it; and that it was hard for him to believe I made it from cane grown in

Newton. I may add in another place letters from gentlemen who tested the article made by me, and shall give testimony from other states in regard to syrup produced there.

A great deal of alcohol can be made from the juice of this plant, whether grown North or South; and it is certainly worthy of attention on this account, as alcohol is used to a great extent for mechanical and other important and proper purposes. In regard to cider, or the champagne-like drink the juice, when fermented, yields, and the facility with which it can be furnished, I shall have but little to say; for I very much doubt if any practical good would follow its introduction. But it is a fact that it does yield such a drink.

Of the juice as a coloring matter, when set with alum, I can only say, from *experience*, that I did not fully succeed, the dye not coming up to my expectations in brilliancy; and though much is claimed for it by others for such purposes, still I am not inclined to estimate it very highly.

Of the waste, or "begass," which has been heretofore referred to, I cannot speak from experience; but I have no doubt it will prove all that is claimed for it as a substance from which good paper can be made. It seems to possess the fibre sufficient for such a purpose, and will probably be fully tested another season. If it should so prove, it will fill an important place in the manufacture of this article.

And, then, the seed which it yields so profusely possesses all the rich qualities of rice, or other grain, to feed out to cattle, swine, or fowls. It would seem to be almost worth growing for that alone, as it yields from twenty-five to fifty bushels per acre. And, lastly, the fodder, which must be quite valuable, on account of its containing so much saccharine matter. It may be sown for fodder, like corn, and will give two crops; for, unless the season is quite dry, it will quickly shoot up again after being cut down. Or, where grown for sugar, or syrup, the leaves and tops of the stalks that are too green to be used for sugar-making can be saved for fodder, and thus no part of the plant be lost. If desired, the brush-top may be used for making brooms. Especially would I recommend its trial as a green crop for soiling, or for curing, for winter feed for cattle; for I think it will prove far superior to any and all crops that are now grown for that purpose.

I now propose to give the particulars of my

own experience during the past season with this truly wonderful plant. I received the seed from the Patent Office, through my friend, Hon. Simon Brown, editor of the *New England Farmer*, and, believing it to be a *humbug*, I planted it about the twentieth of May, in hills two feet apart, the rows three and a half, manured in hills as for Indian corn and no more, on a dry, gravelly soil, covering the seed lightly, — for if covered too deep the seed decays. In a few days it made its appearance, resembling corn, or more like broom-corn, or barn-grass, and would be mistaken by the ignorant for that grass, and there would be danger of destroying it when hoeing. After it had been up about ten days, I had it hoed, and treated it all through the summer as I treated my corn. When the panicles made their appearance, which they did about the first of September, I cut them off of all that which I intended for sugar or syrup making, while that which was intended for seed was left until just before the frosts came, when it was cut up and spread in a dry place. Most of the seed ripened, though it was planted late, and the season was cold and wet, and for weeks in the spring and early summer the plants made little growth.

The cane attained the height of ten and a half feet.

I tried my first experiment with the cane the last of September, and found the juice was thin and less rich than at a later period. After expressing the juice, which is of a light green color and nearly as thin as milk, I put it immediately over a slow fire, without putting anything into it to clarify it. As it gradually warmed, I removed the green scum that rose on the surface, until it boiled, and there was no further need of skimming. I let it boil until four fifths had evaporated, and then turned it off to cool. The result was a very nice syrup.

In the second experiment I took the cane about the tenth of October, and expressed the juice as before, putting it over a slow fire and gradually raising to the boiling point, this time putting in a small quantity of lime-water, both to aid in purifying and to neutralize the acid which the juice contains. This time the juice appeared and proved much richer than before. The same process of skimming was carried on, and I obtained a much larger proportion of syrup and of a better quality, such as I have described on a former page. Subsequent to this I tried another experiment with the juice, and proceeded as before, except I boiled it more, and then set it away in a strainer to drain ; and it grained tolerably well, though the sugar was of rather dark color. I tried the juice for coloring, as I have before said, with indifferent success. In addition to this, I used the seed-cane stalks and leaves for fodder ; cut up the stalks and fed to horses, cows, and swine, and they would eat it with the greatest avidity, even like shelled corn. This ended my experiments with the cane.

I now propose to give the results of experiments that have been tried by others in different parts of the country. And first among them stands RICHARD PETERS, Esq., of Atlanta, Georgia, who has tested it more fully than any other man, so far as I know, in the United States. He says: "I considered it a 'humbug' until my children, towards fall, made the discovery of its being to their taste equal to the true sugar-cane. This year I planted one patch April fifteenth, another May eighteenth, on land that would produce, during a 'seasonable' year, forty bushels of corn per acre, and this year not over twenty bushels. Seed sown carelessly in drills, three feet apart, covered with a one-horse plough ; intending to 'chop out' to a stand of one stalk six inches apart in the row, but failed to get a good stand, as the seed came up badly, from the deep and irregular covering. Worked out same as for corn, ploughing twice and hoeing once. I determined to give the syrup-making a fair trial; consequently ordered a very complete horse-power mill, with vertical iron rollers, that has worked admirably, crushing out juice for eight gallons of syrup per hour, worked by two mules, with one hand to put in the cane and a boy to drive. On the thirteenth of September, finding the seed fully ripe, I had the fodder pulled and the seed-heads cut. Yield of fodder per acre, eleven hundred to thirteen hundred pounds. Yield of seed per acre, twenty-five bushels of thirty-six pounds to the bushel. First trial of mill, seventy average canes gave twenty quarts of juice; six hundred and six average canes, passed once through the rollers, gave thirty-eight gallons one quart of juice; passed a second time through, gave two gallons of juice. The forty gallons one quart gave eight gallons of thick syrup. I carefully measured an eighth of an acre having the best canes and the best stand, another eighth having the poorest canes and the poorest stand; the result I give below, the canes having passed once through the rollers:

Best eighth of an acre.

Yield of juice from 3315 canes, ...	253 gal.
Yield of syrup from 253 gallons of juice, . . .	58½ "
Rates of syrup per acre, ...	468 "

Poorest eighth of an acre.

Yield of juice from 2550 canes, ...	179 gal.
Yield of syrup from 179 gallons of juice, . . .	43¼ "
Rates per acre of syrup, ...	346 "
Weight of 30 selected canes, ...	49½ lbs.
" of juice pressed out, ...	25¾ "
" of crushed cane, ...	23 "
Loss in crushing, ...	¾ "
Weight of crushed cane dried in sun, ...	9½ "

"The following tests were made at the mill by Dr. Robert Batty:

Specific gravity of Chinese Sugar-cane juice, .	1.085
" " of syrup, ...	1.335
" " of New Orleans syrup, . .	1.321
Thermometer applied to syrup, ...	77 deg.
Thermometer applied to juice, ...	70 "
Saccharometer, " "	25½ "

"The juice should be placed in the boilers immediately on being pressed out, then boil slowly until the green scum ceases to rise; then stir in a tea-spoonful of air-slacked lime to five gallons of juice; continue skimming and boiling until the syrup thickens and hangs down in flakes on the rim of the dipper. I have made the clearest syrup by simply boiling and skimming, without lime or other clarifiers. The lime is requisite to neutralize a portion of the acid in the juice: the true proportion must be determined by well-conducted experiments. The cost of making the syrup in upper Georgia, in my opinion, will not exceed ten to fifteen cents per gallon. This I shall be able to test, another season, by planting and working up fifty acres of the cane. I am satisfied that this plant will enable every farmer and planter in the Southern States to make at home all the syrup required for family use ; and I believe our chemists will soon teach us how to convert the syrup into sugar for export, as one of the staples of our favored clime. Obtaining such unlooked for success with the Chinese sugar-cane, I concluded to try our common corn. From a 'new ground,' planted three feet by three, one stalk to a hill, a week beyond the roasting stage, I selected thirty stalks.

Weight of 30 stalks,	35¾	pounds.
" " juice,	15¼	"
" " crushed stalks,	19½	"
Loss in crushing,	½	"
Yield of syrup,	1½	pints.

The syrup was of a peculiarly disagreeable taste, entirely unfit for table use."

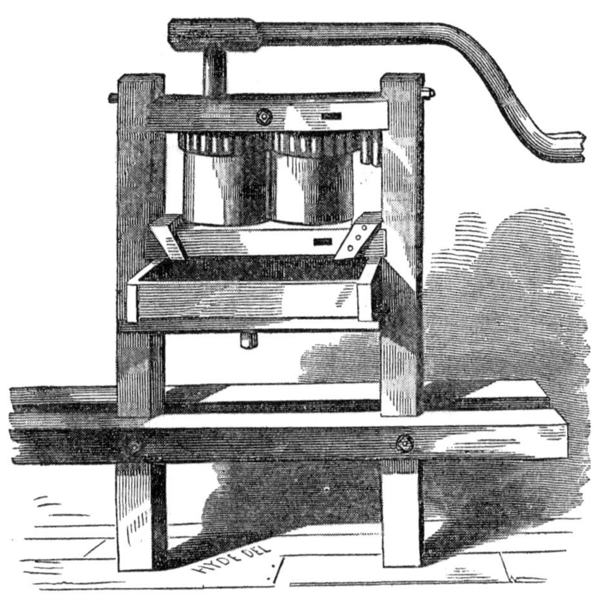

The mill referred to in the above communication, and of which I give an engraving, I understand was made at Atlanta, Ga., and cost for iron work about forty-five dollars; and is said, by a committee of gentlemen who examined it, to be "worthy of commendation." This committee further say, that they "have no hesitation in pronouncing upon the value of the Chinese sugar-cane for making syrup." Without asking anybody to embark largely in the cultivation of this new article, I think it well worth a fair trial, "and hope that none will be ready to write it a humbug until they shall have tested its merits."

"We have repeatedly called the attention of our readers," says the Charleston *Mercury*, "to the value of the Chinese sugar-cane, and are therefore greatly rejoiced to find that the article has fallen into the hands of so scientific and careful an experimenter as Ex-Gov. Hammond, who will be widely recognized as one of the highest authorities in Southern agriculture. We copy below a carefully considered report of his experiments with the Chinese sugar-cane, prepared for an agricultural society in his own neighborhood, and furnished for publication to the Barnwell *Sentinel*. It will commend itself to the attention of the planters of the state. There can be no doubt that the sugar-millet is destined to prove an important addition to the resources and comforts of the plantation."

" *Report of an experiment in making syrup from the Chinese sugar-cane, or sugar-millet, made to the Beech Island Farmer's Club, August 2d,* 1856. *By Hon. J. H. Hammond, of South Carolina.*

"One of our members, Mr. Redmond, of the *Southern Cultivator*, distributed among us, last winter, some seeds of what is commonly called sugar-millet. He very kindly gave me enough to plant half an acre,—about a pint. I prepared a plot of ground on a northern slope of old, stiff, worn-out land, in such a manner and with so much manure as would probably have made it yield, with average seasons, about twenty bushels of corn per acre. On the 22d of March I planted the millet-seed in three-feet drills, dropping every eighteen or twenty inches some six or eight seeds. It was ploughed and hoed often enough to keep the grass down, and about the first of July began to head. The heat had then been unusually intense for two weeks, and has continued so up to the present time; and latterly the drouth has been very destructive. I do not think this half-acre would have yielded five bushels had it been planted in corn. Having intended, however, to ascertain whether the millet would make syrup, I had a rude mill put up, with two beech rollers. Finding by the 22d of July the most advanced heads had passed the milk stage, I had 1750 canes cut, that I supposed were a fair sample of the patch. The first three or four hundred were passed through the mill twice, the remainder four times; and the yield was 194 quarts of juice. But ten canes that I selected and passed seven times through the mill yielded three quarts. The juice was received into common tubs, and tested by a thermometer, and a saccharometer with a scale of 40 degrees. The thermometer stood in every instance at 78 degrees. The saccharometer varied from 21½ degrees to 23½ degrees. At the latter point the juice would float a fresh egg. I boiled it in a deep pot, and after six to seven hours' boiling obtained 32 quarts of tolerable syrup. The next day I selected ten canes, the heads of which were fully matured; ten more, in full milk; ten more, the heads of which were just fully developed and the top seed beginning to turn black; and again ten comprising all these stages, but from which I did not strip the leaves. They were all passed through the mill seven times, and yielded nearly the same quantity of juice — about three quarts for every ten canes. The juice — tested by the saccharometer — showed that the youngest cane had rather the most, and the oldest rather the least saccharine matter. The whole together, with that of a few other good canes, exhibited at 80 degrees of the thermometer 24½ degrees of the saccharometer. From forty-two pints of the juice I obtained, after four hours' boiling, nine pints of rather better syrup than that made the day before. In these boilings I mixed with the cold juice a tea-spoonful of lime-water, of the consistency of cream, for every five gallons. These selected canes grew on the best spot on the patch, and where corn probably might have been produced the present season at the rate of twenty bushels to the acre. They were one inch in diameter at the largest end, and seven and a half feet long after cutting off the head and the foot of the stem. After this I cut down all the inferior cane, and cured it for forage.

"On the 28th of July, two of the members of the club, being at my house, remained to see the result of pressing and boiling four hundred canes I had cut and stripped. Each of us selected ten canes, and put them through the press eight times—the result being as before, about three quarts for every ten canes. But even after the pressure juice could be wrung from the canes by the hand, and we agreed that at least one fourth of it, and that the best, remained in the cane—so inefficient was my mill. The rest of the cane I ordered should be pressed six times; but we did not ourselves remain to see it done, nor did we count the 400 canes. The yield of the whole, however, was thirty-seven and one half quarts, with the thermometer at 85 degrees in the juice: the saccharometer stood 24½ degrees. We boiled the juice until it run together on the rim of the ladle, and hung in a transparent sheet half an inch below it before falling, and this in two and a half hours. The result was six quarts of choice syrup. The next day I repeated the experiment on a larger scale, with equal success; and I have brought to the club enough of the syrup to enable every member to try it, and judge of its quality. All who have tested it agree that it is equal to the best that we get from New Orleans. In these last boilings I put a table-spoonful of lime-water, prepared as before, to every ten gallons. The whole process of clarifying and boiling was carried through in the same pot, and that very unsuitable from its depth. I measured the grain from a number of heads, and the result was an average of a gill from each. I weighed a half a peck of maturer grain after several days' exposure to the sun; — it weighed four and three fourths pounds, equal to thirty-eight pounds per bushel. I weighed twenty of the best cane cut for forage, after it was cured sufficiently to house. They weighed twenty-four pounds, equal to thirty thousand pounds for twenty-five thousand canes; which I think might be grown on land that would make twenty-five bushels of corn, with average seasons. I have tried horses, cattle, and hogs, and find they eat the cane, its leaves and seeds, greedily, and fowls and pigeons the last. I think, however, that, when allowed to mature, the cane should be cut up fine for animals, as the outer coat is hard. I did not attempt to make sugar, not having prepared for that; there can, however, be no doubt that sugar can be made from such syrup as this. And as they make more syrup in the West Indies per acre than they do in Louisiana, only because the cane matures better, it is not unreasonable to infer that the millet, which matures here perfectly, and would even make two crops in one year, will yield more and better sugar than the Louisiana cane.

"Beginning to cut cane as soon as the head is fully developed, it may be cut for a month before it will all ripen,—how long after that, I do not know. As succession of crops might be easily arranged so as to insure cutting and boiling, from the first of July,—probably earlier,—then until frost, I have housed some stalks immediately from the field, to ascertain, hereafter, whether thus treated it will yield juice and make syrup next winter. A good sugar-mill, with three wooden rollers, may be erected for less than twenty-five dollars, and a sugar-boiler that will make thirty gallons of syrup a day may be purchased in Augusta for less than sixty dollars. This millet will, of course, mix with any other variety of the millet family planted near it. I have now stated the chief particulars of my experiment. A single experiment — especially one in agriculture — is rarely conclusive. I may err myself, and cause others to err, were I to express with any emphasis the opinion I entertain of the value of this recently-introduced plant."

We learn by this experiment, though tried at the South, much that is of importance to those who are entering upon the cultivation of this sugar-cane.

Among those things, we propose to notice briefly, first, the quantity of seed used to "half an acre,— about a pint." This, according to my experience, is a small pattern, though I have no doubt, if it were evenly distributed with a seed-sower, it might answer. But it is always better to plant more than you want, and thin out, than to plant so thin that you will fail to get a crop. We learn further that it was planted in drills, one seed every three inches. I believe this to be the best way to raise it — in drills — either for syrup-making or for fodder, though care should be taken to thin out, if too thick, so that the cane would be stout enough to resist the storms that sometimes lay the stover corn prostrate. We also find that that which was planted the 22d of March was fit for cutting on the 22d of July, being a period of one hundred and twenty days, which accords with my experience with it during the last season, and which shows most conclusively that it can be grown in the New England States; for it may be put into the ground ordinarily as early as the first to the tenth of May, and consequently would be suitable for cutting from the first to the tenth of September. This would give us the whole month of September, and, in some seasons, considerable of October, in which to manufacture our sugar, or syrup. Again, we learn from the account that it withstood a most severe drouth, which it is said would have proved very severe to corn, and probably materially lessened the crop, and yet the cane did not suffer much. We are also shown the manner in which the juice was clarified, which we shall do well to notice, for I believe there is no better clarifier than lime-water, though there needs to be some careful experiments to determine the quantity that shall be used. For, where

I used the lime my syrup was very much better than that made without the lime. We are further told, in regard to the seed or grain, the amount from each head being a gill, and weighing at the rate of thirty-eight pounds to the bushel ; which goes to confirm the position I took in a former part of this work, that it would almost, if not quite, pay for raising for the grain. I also agree with the writer in saying that the stalks should be cut up with a hay or stalk cutter, or in some other way ; for the stalk is hard, and cannot be so readily eaten as when cut. Again, it is said that a sugar-mill, with wooden rollers, can be procured "for less than twenty-five dollars, and a sugar-boiler that will make thirty gallons of syrup per day for sixty dollars." Now, it is doubtful if such a wooden mill will be a proper one, such as it would be economy to use, even though it should seem to answer the purpose tolerably well. Sugar-mills, such as are manufactured to send to the West Indies, are made at South Boston, and all complete cost from three to five hundred dollars ; though, if this cane should succeed, I am in hopes we shall have cheap, portable mills, one of which might answer for a small neighborhood. Boilers can probably be obtained cheaper here than at the South, though it is not necessary to have these large boilers except when it is intended to make large quantities of syrup.

We will give further results of experiments made at the South, and quote from the *Southern Cultivator*, for October, 1856: "In the winter of 1844–5, the junior editor of this journal obtained from Boston a few ounces of seed of this plant, — Chinese sugar-cane, — then newly imported from France. It came very highly recommended as a sugar-producing and forage plant; but, having a vivid recollection of many previous disappointments with new-fangled notions, we concluded to test it cautiously and moderately. In order, however, to give it a fair chance, we distributed small parcels, per mail, to friends in various portions of Georgia and the adjoining states, and planted for ourselves only seven or eight hills, in a poor spot in our garden. At first it came up like grass, or Egyptian millet, and grew off slowly and weakly; but in a few weeks it began to shoot upward, and in less than three months attained the height of eight or ten feet, with large and well-filled heads of seeds, somewhat resembling broom-corn, but covered with a black husk, or chaff. Passing by it one day, when the seeds were nearly or quite ripe, we concluded to test the sweetness of the stalk; so, cutting a moderate-sized cane, and peeling its hard outside coat, we found a solid pith of about three fourths of an inch in diameter, and crisp, brittle, and an exceedingly sweet and pleasant flavor, wholly and entirely unlike anything of the corn-stalk family that we had ever tasted. It was, in fact, *ready-made candy;* and as soon as the younger members of the family and the negroes got the taste of it, we were obliged to interdict its further use, in order to save seed. When the latter were fully ripe, we cut off the heads and saved them carefully, noticing, with some surprise, that the leaves or blades of fodder were still as fresh, green, and succulent, as ever. The stalks were then cut off near the ground, and fed, leaves and all, to our horses, mules, and milch cows, all of which eat of it with the greatest apparent relish and avidity. Considering that crop disposed of for the season, we paid no more attention to the stubble, or stumps, until we happened to notice that, millet-like, they were shooting out anew, and pushing on for a second growth. This growth we watched with some interest, until the first frost checked it; at which time, the stalks were six feet high, full of broad and juicy leaves,

5

and with the second crop of seed just making its appearance above the 'boot.' Fully satisfied by this time that it was valuable, at least for the production of soiling, forage, and dried fodder, we next turned our attention to its saccharine properties, and fortunately induced our friend, Dr. Robert Battey, of Rome, Georgia, who was at that time pursuing the study of experimental chemistry, in the well-known laboratory of Prof. Booth, of Philadelphia, to test it. As the result of his experiment, Dr. Battey sent us three small phials, one containing a fine syrup, one a sample of crude brown sugar, and the other a very good sample of crystallized sugar. This we believe to be the first crystallized sugar made in the United States from the juice of the *sorgho-sucré;* and as Dr. Battey's opinion of its value as a plant fully agreed with the reports of the French *savans* who had investigated its properties, and with our own convictions, we this year disseminated it more widely, and planted nearly two acres, for the express purpose of raising the seed, and testing the ability of the plant to bear repeated cuttings, like Egyptian and other varieties of millet. It was planted very late, on poor soil, and has received but imperfect culture; and yet, at the present time (August 25) a portion of it has been cut *three times*, and is growing up finely, while the remainder has ripened its seed, and will yield a full crop of excellent fodder after the present stalks are cut off at the ground and crushed for syrup, or fed out to our stock.

"So much for its introduction into this section, and its history among us thus far. It is our deliberate opinion, that for 'soiling,'— cutting green, repeatedly, — for the production of syrup, sugar, cider or wine, alcohol, fodder, and grain, at the same time, it will be found invaluable to the South, and that no plant of recent introduction among us can at all compare with it." It will be seen by the above that the cane will produce, as I supposed, more than one crop of fodder from the same roots, even in the Northern States, and three or four in the Southern. In this, as in other respects, it may prove more valuable to the South than the North. But, after giving some more testimony from the South, we shall give some from the more Northern States.

Dr. Battey, to whom reference has been made on a former page, writes as follows to the *Southern Cultivator*, in reply to inquiries that had been made of him: "I cheerfully comply with your request for information on the subject of my observations and experiments on the

Chinese sugar-cane, as a syrup-producing plant. My attention was first called to the subject by the seed which you were kind enough to send me in the spring of 1855. I planted them, and raised, say, fifteen to twenty canes, that year, from which I extracted a small quantity of juice for analysis. This juice, as you are already aware, yielded, during the winter, sugar and syrup, samples of which I sent to you for inspection. Impressed as I was with the probable importance of this plant to the agriculturists of the South, I did not deem it prudent to speak hastily of its merits, waiting, rather, until a repetition of these experiments upon a larger scale should fully establish the opinions I had entertained of it.

"The present year I have cultivated a few more canes for my experiments, and upon the farm of Richard Peters, Esq., Gordon county, Georgia, I have witnessed the growth of the cane by the acre, and the production of the syrup by barrels. I have, in the mean time, read attentively the opinions of Gov. Hammond, of South Carolina, and others in different sections of the Union, who have grown the plant and experimented with it, as also the valuable paper of M. Vilmorin, of France, who has given this subject much study and investigation; so that, calmly viewing all the facts which I have been able to collect, I no longer entertain a doubt that this plant is well worthy of the attention and study of the farmers and planters of the South. If the opinions I shall express should seem to some too wild and extravagant, I trust they will receive them as the honest and candid sentiments of one who has carefully examined the subject, and be led to investigate and experiment for themselves. Should I thus be enabled to arouse the attention of Southern farmers to the importance of this plant, my object will have been accomplished, and my labor well expended.

"The Chinese sugar-cane seems to adapt itself to all the vicissitudes of our varied climate and soil, and with a facility unsurpassed by corn or wheat. In Cherokee, Ga., it flourishes in a high degree of perfection upon soil high and low, rich and comparatively poor, producing heavy crops of stalk, leaf, and seed. The experiments of Mr. Peters (which are already published in many of our agricultural papers) present an example of most successful culture. I have found it to grow with me, in all respects, as vigorously as corn, with precisely similar treatment. In Alleghany County, Md., a correspondent writes for the May number of the *American Farmer:*

"'I think it well adapted to our mountainous country, and promises to be more valuable than any other article we can grow for provender. I believe it will produce six or eight tons of dried provender to the acre.' The present writer has met many intelligent and enterprising farmers of Pennsylvania, Maryland, Virginia, New Jersey, and New York, in attendance at the late National Fair at Philadelphia. Many of them had witnessed its growth in their respective states with entire success. One gentleman of New Jersey had grown a half-acre of the cane this season. It has been successfully grown in Illinois, also; and one gallon of the juice is said to have yielded, by boiling, a quart of syrup of good quality. There is every reason to conclude that the cane may be easily and successfully grown *in all parts of our country.*

"CULTURE. — While the seed remains in the hands of the few, and commands a price too high to permit a waste, it should be planted for one season with good distance, that the seed crop as well as the cane may attain their highest state of development. I would recommend that the rows should be three, or even four feet apart, and the distance of, say, two feet given in the row, dropping one or two seed in a place. Let the ground be well cultivated, as for corn, and the shoots or suckers which spring up from the root be all permitted to grow. A small portion of the crop should be reserved for seed, and permitted to stand until fully matured and dry. It would be well to limit the canes in the seed patch to one. By all means permit no *broom-corn, Dourah-corn*, or other plants of the same family, *to grow near your cane.* It readily intermixes with these varieties, and effectually ruins your seed for the production of syrup. For the same reason, great care should be observed in procuring reliable seed, as well as in keeping them so.

"After the first season, when a full supply of seed shall have been secured, a better-paying syrup crop may be grown by closer planting. The space between the rows may well be narrowed down to three feet, and the seed put in, say, two or three every six inches; when well up, the stoutest and healthiest plants should alone be allowed to stand. The cane, when very young, presents so much the appearance of grass, that an advantage may perhaps be gained by dropping some other seed with the cane, that the latter may be more readily distinguished. This, of course, should be drawn out with the superfluous cane-plants. When of sufficient size, the plants should be suckered down to one cane for each root. In other respects, the successful grower of corn will not be at a loss in the cultivation of this plant. I have found a suitable time for planting to be immediately after the corn crop, although excellent results have been obtained by planting as late as the 15th of May, in Cherokee, Ga. It will doubtless be desirable to make several successive plantings, that they may mature gradually, and so give more time for harvesting the crop. The land, in my opinion, should be prepared in all respects as for corn.

"HARVESTING. — When the stalk shall have attained its full size, and the seed have passed from the dough stage to a harder texture, the cane may be considered sufficiently mature; or, if the crop be large, and a deficiency of hands be apprehended, the cane may be cut earlier, and the cuttings continued from time to time, as needed for the press. The fodder should be pulled as for corn; another set of hands cutting off one half to two feet of the top with the seed, while others cut the cane at the ground and throw it into piles, from whence it is handed to the press. Prior to the harvesting, a set of proper rollers and kettles should be provided, and well set up ready for service.

"The mill made use of by Mr. Peters, and which was gotten up under his direction for the purpose, is, in my opinion, of very unexceptionable quality for a small apparatus, and works admirably. It is of a suitable size for a small crop, and no farmer should undertake to supply its place by wooden rollers for a crop of even two acres. The loss of juice will more than counterbalance the difference in expense. It is worked by two mules. Three kettles, of from sixty to one hundred gallons' capacity, will be required to keep pace fully with the mill; it is desirable that these should be broad and shallow, that they may present a large evaporating surface, and substantially set in brick for security and convenience. They should not be distant from the press, and if upon ground lower than the latter, an advantage is gained in running the expressed juice directly into them, and thus saving the labor of transfer.

"PRESSING.—The canes, located conveniently at hand, are one by one doubled in the middle and forced between the rollers, which are kept in as close proximity as the strength of the mill and the power of the mules will warrant. An active hand will feed the mill easily, if the canes be placed within his reach. A boy is required to drive, and if the mill be well constructed to throw off the begass from behind, nothing more is required except an occasional removal of the latter by a pitchfork, to keep it out of the way of the mules.

"BOILING DOWN.—One of the first things done, in commencing operations, should be to start the fire under the kettles, that they may be well warmed by the time the juice is ready for them. The fires should be so arranged that they may be under good control, to be forced or withdrawn as occasion may require. When the juice is placed in the boiler, the fire should be gradually increased to a simmering heat, "not to active boiling," and maintained at this temperature until a thick green scum rises to the surface and forms into puffs, seeming ready to crack. This scum, when fully formed, should be removed clean from the surface. The heat may now be raised to boiling, and kept in an active state of ebullition, until the bulk is reduced one half. The fire may now be removed from one kettle, and its contents be transferred to the other, when the heat must be gradually moderated as the syrup becomes more concentrated, to avoid the danger of scorching, which injures the color and flavor. Should more dirty-green scum rise to the surface after the first skimming, it should likewise be removed.

"In regard to the precise degree of concentration to which the syrup should be brought, it is exceedingly difficult to lay down any precise and simple rule, which shall meet every case. The plan for determining it in use on the sugar plantations, and which was adopted by Gov. Hammond and Mr. Peters, is based upon the judgment of the eye in respect to the consistence of the syrup when poured from the ladle and cooled as it drops from its edge. This test is evidently very defective, since the temperature of the atmosphere regulates the consistence which the syrup must assume on cooling down; so that a syrup boiled on a cold day will necessarily be thin and watery as the weather moderates, and a syrup finished at night will differ materially from that of the noonday. Although a good approximation, it is not exact enough for the tyro to secure a desirable uniformity in the consistence and value of the product, or to obviate the danger of fermentation and loss. To remedy this uncertainty, and secure a uniform result at all times, I have constructed a simple instrument, which determines readily and with certainty the precise moment when the syrup should be removed from the fire and transferred to the barrels. For the convenience of those who may desire this aid, I shall prepare a number of them during the season, which may be furnished by mail. With such a guide to the uninitiated, there are certainly few more simple operations upon the farm than the manufacture of syrup from this cane.

"It is a prevalent opinion that lime should always be added to the juice as soon as it is pressed out, and the idea has been advanced that it could not be clarified without lime. This is undoubtedly a mistake; the juice alone, under my hands, clarifies itself more readily without lime than with it. The latter answers no useful purpose, as far as the syrup is concerned, save to neutralize the free acid (phosphoric) which exists naturally in the cane. Lime darkens the color, and, to my taste, detracts from the peculiar grateful flavor of the syrup. Many would, perhaps, object to the slight acidity; to such I would say, use the lime, but use it sparingly. To prepare it for use, take a half-peck of lime, slake it in a bucket of water gradually added, stir up well, and strain the milk through a cloth ; let it settle for half a day, pour off the water, and dry the powder. Of the latter you may use from a half a tea-spoonful to two tea-spoonsful for every five gallons of juice, after the scum has been removed.

"The scum is used in the West Indies for the manufacture of rum, the details of which are entirely too elaborate to be introduced here. It may be also advantageously disposed of as food for hogs. The quantity of saccharine matter left in the begass renders it a nutritious food for stock. This refuse, by leaching water through it, yields a saccharine solution which may be fermented into beer or vinegar, and may be distilled into whiskey and alcohol. It may be also advantageously used to cover the cut canes in hot weather, when it may be desired to have a large quantity kept at the mill for days and weeks before being all used. The constant evaporation of the juice in the begass keeps the cane beneath at a temperature so low as to prevent fermentation, as well as the drying of the cane ; it will also serve to shield it from the frost. A suggestion has been made to convert the ligneous fibre into paper. It certainly is a better material for this purpose than much that is now employed. It is, however, an object of minor importance to the Southern planter as yet. As a manure, the begass is evidently a most valuable article, for its large amount of phosphoric acid, added to the decomposing vegetable and the other mineral matters which it contains, while the remaining portions of saccharine juice readily induce a fermentation which ends in putrefaction, and leaves the mass in a fit state for the nourishment of plants. The large quantities of mineral matter, and particularly the phosphoric acid, which the cane in its growth must remove from the soil, necessarily imply that it will be an exhausting crop, since these materials certainly cannot be furnished by the atmosphere. This evil may, in great part, be removed by carefully returning to the soil again the refuse in form of manure. If other fertilizers be needed to repair the waste, Mexican phosphatic guanos, which are now offered at low prices, would doubtless be advantageous.

"In the experiments by me, during the winter of 1855, and also at the farm of Mr. Peters, in September last, I was forcibly struck with the better quality of the juice grown in our section of country, as compared with that experimented upon by Mons. Vilmorin, whose paper will be found translated for the present year's *Working Farmer.* He gives the density of his sap at 1.050 to 1.075, while that examined by myself was uniformly found to be 1.085, with but little variation, and in every case some small corrections for temperature, which would increase the specific gravity slightly. The average density given by various observers in the West Indies, of juice from the several varieties of sugar-cane grown in these colonies, is about 9 degrees Baume, corresponding to a specific gravity of 1.064, — less, considerably, than mine. From this fact, however, it is not to be inferred that the juice of our cane abounds more largely in saccharine matter than that of the West Indies; for such probably is not the fact; for the former is known to contain a larger proportion of salt and vegetable matters than the latter. It argues only the remarkable adaptation of the Chinese cane to our climate and soil. M. Vilmorin obtained from this 'sap' of the densities named, from 1.050 to 1.075, on the

13th of October, 1853,	10.04	per cent. saccharine matter.
28th of November, "	13.08	" " "
28th " (2nd trial)	14.06	" " "
14th " 1854,	16.00	" " "

Of the latter, 11.75 were uncrystallizable, and

but 4.25 of the crystallizable variety. M. Avequin obtained from the juice of this cane, grown, I presume, in Louisiana,

Saccharine matter, ...	152
Salts and organic matter, . . .	10
Water, ...	838
Cane-juice employed, ...	1000

I have not been able to compare these experiments with similar results obtained here. I propose doing so the coming season. M. Vilmorin estimates the percentage of weight of juice obtained by him at 50 to 60 parts in the 100 of cane employed, and remarks that even 70 per cent. can be easily obtained by proper machinery. Mr. Peters obtained from his mill an average of 50 per cent., and juice could be readily wrung from the begass by hand. Thirty canes were sorted out and weighed by myself, and, after grinding, gave the following results :

Thirty canes weighed, ...	52	lbs.	14	ounces.
Juice collected, ...	26	"	1	"
Begass, ...	26	"	7	"
Juice lost in mill, say, . . .			6	"

The juice actually extracted weighed precisely one half that of the cane used. Two pounds of the begass was weighed and carefully dried, and gave twelve ounces, showing a loss of one pound and four ounces of water, which represents $21\frac{7}{10}$ ounces of juice; so that the quantity of juice remaining behind in the begass may be put down at seventeen pounds, fifteen ounces. The result now stands,

Juice collected,	26 pounds,	1 ounce,	or	49.30 per cent.
" lost in mill,		6 "	or	.70 "
" in begass,	17 "	15 "	or	34.05 "
Woody fibre,	8 "	8 "	or	15.05 "
Cane used, 52 pounds, 14 ounces.				100 per cent.

In other words, we have $84\frac{1}{2}$ per cent. of juice, and $15\frac{1}{2}$ per cent. of woody fibre. From these figures it would seem that 70 per cent. in juice ought to be easily obtainable by proper machinery, and it becomes more apparent when we take into consideration the soft, compressible texture of this cane as compared to that of the West Indies. Mr. Peters states the yield of his best $\frac{1}{8}$ acre in syrup at $58\frac{1}{2}$ gallons; that of the poorest $\frac{1}{8}$ acre at $43\frac{1}{4}$ gallons. Taking the average, we have, as the yield of the entire acre, 407 gallons; assuming the yield of the juice to correspond with the average results obtained by experiment, say 50 per cent. of the entire weight, with proper machinery, expressing 70 per cent., we have a yield of 570 *gallons per acre.* I examined carefully the specimens of syrup boiled under the eye of Mr. Peters, and also by myself. Several of these specimens were of a superior quality, all of them surpassing my expectations, in view of the crude manner in which they were made. There is present in all of them, to a greater or less degree (owing to differences in manipulation), a peculiar flavor, reminding one of the maple-sugar, which is very grateful to the palate, and gives it a decided preference over the article which we get under the name of New Orleans syrup. This, so far as I know, has been the uniform judgment of all who have tasted it. These syrups give a precipitate of foreign matters with the basic acetate of lead (a delicate test), little, if at all, greater in amount than the New Orleans syrup. The precise nature of these precipitates remains to be ascertained and compared. The syrups vary considerably in density: those from the Chinese cane ranging from 1.298 to 1.335, while that of the New Orleans sample was 1.321. This variation in the density is an evil which should be corrected, to produce a good marketable syrup, which shall keep well. Samples of the Chinese cane syrup have been valued by the intelligent dealers in the article, in our section, at from 65 to 75 cents the gallon, by the barrel.

"In calculating the yield of this crop, we must take into consideration twelve hundred pounds of excellent fodder, and twenty-five bushels of corn, worth, as food for stock, say two thirds the value of the ordinary corn ; so that we can fairly offset against the syrup crop, in the way of expenses, nothing more than the labor of its manufacture, for the forage and corn will repay the expenses of the culture. A full consideration of the facts, which have been passed over somewhat in detail, must make it evident to the mind of every intelligent farmer that this plant presents, at the present time, a *promise of reward* for its culture unequalled by any which has been introduced upon our soil since the introduction of the cotton crop."

The above information is of a very valuable character, such as I have not been able to get from any other source, and such as can be depended upon. It gives us rules for planting, harvesting, and manufacturing, which are, for the most part, applicable to any and all latitudes where the cane will grow. In regard to the use of lime, I would say that I should much prefer to use it, even if it did not assist in clarifying; for we know it will neutralize the acid which the juice contains, and so make it more pleasant to the taste, and tend to preserve it longer. The syrup alluded to, which Mr. Peters made, I have seen; and noticed that it was of a darker color, and not so heavy as that which I made, for the reason it was not boiled so much. Still, the syrup is such as would sell readily at the prices named by Prof. Battey. I add the following from the *National Intelligencer*:

"The Chinese Sugar-cane has come to be the ordinary name of the Sorgho Sucré, a most valuable plant of the sugar-cane order, and, therefore, allied to the maize or Indian corn, but more nearly to the broom-corn. Its cultivation has commenced amongst us, and there is now in Washington more than an acre of it, growing luxuriantly, and promising a yield of considerably upwards of one hundred bushels of seed, besides many tons of stems and foliage, rich with saccharine fluid and solid food, material for horses, neat cattle, and swine. Not only here, but in various and widely-distant parts of the Union, has trial been made of it, and with uniform gratifying results. We have read a letter from a farmer in Illinois, who has tested its character, and reports of it in the most favorable manner. Out of a gallon of the liquid sap in the stem, which he expressed by the primitive contrivance of a rolling-pin, he obtained, by boiling, a quart of molasses, with very little impurity, and of approved taste. The usual proportions of sugar to sap lie between fifteen and twenty per cent., the crystallizable sugar increasing with the decrease of the latitude. Beside this proportion of sugar, there is an amount of perhaps five or eight per cent. of uncrystallizable sap, from which a very agreeable beverage can be made, and alcohol distilled more cheaply than by any other method. This sap, strange to say, if set with the oxide of tin, will dye silk of a beautiful pink. As a food-plant for stock of all kinds, it seems to overtop all we now possess, furnishing, in fair soils, twenty-five tons per acre of excellent fodder, every bit of which is greedily eaten by animals. The seeds, too, by which the plant is propagated — in this, unlike and superior to the sugar-cane of Louisiana, which is raised by cuttings — are fit for human food. At all events, when ground and made up into cakes, after the manner of linseed cakes, they supply a good material for fattening stock. The brush, or top from which these seeds are taken, is not without its service ; for the plant is a species of broom-corn, and, therefore, its top, when deprived of seed, answers well wherewith to manufacture brooms. When the sap, top, seeds, and leaves, are taken, leaving only the crushed stem, it still has an economic value, for paper can be manufactured from it.

"This valuable addition to our vegetable productions is originally a native of China, but has been sedulously cultivated for several years in South-eastern Caffraria, whence it passed into France and Algeria, in which last country it comes to great perfection. It would be hard to calculate its value. It constitutes every farm on which it is grown its own sugar-camp, orchard, winery, and granary, as well as a stock-farm and dairy. Indeed, the sorgho may be deemed a sort of vegetable sheep, every part and constituent of which is valuable."

It still further says, in another article: "Among the exotic plants recently introduced into this country by the Patent Office, in the prosecution of its agricultural operations, is the Chinese Sugar-cane, or Sorgho Sucré (*Sorghum Saccharatum*). The history of its introduction, and some account of its success, have been, from time to time, laid before the public through the columns of the *Intelligencer* and other channels, exciting the scepticism of many, and even the derision of some, but, fortunately, awakening the curiosity and enterprise of discerning and intelligent agriculturists in various sections of the United States. We have now the gratification of realizing the happy results of the investigations and labors of this latter class in the successful cultivation, it is hoped and believed, of one of the most valuable products of the soil that has ever engaged the attention of the husbandman,— a product which there are well-grounded reasons for assuming will, of itself, in a brief period, more than recompense all the pecuniary aid and labor that have been bestowed upon the whole subject of agriculture by our government, in the introduction of a plant that may be propagated with advantage in every locality in the Union, that will provide an essential aliment and a luxury to every family at an exceedingly low cost, and that may before long enable us to export to various portions of the world an article of merchandise that we now import to the amount of nearly fifteen millions of dollars a year. It is a singular and gratifying coincidence that the introduction of this plant, and the discovery of its great excellence and adaptation to the soil and climate of many regions of the United States, should be made at the precise moment of the apparent decadence of the culture of the sugar-cane upon the plantations of the South. That this may not result to the disadvantage of the important interests involved in these plantations, is not only desired but believed by those who are fostering the cultivation of the new plant; for it appears to be the accepted opinion, that, though the latter may prosper in any locality in which maize or Indian corn succeeds, yet the soil and climate capable of producing the sugar-cane will prove the best adapted of all to the sorgho sucré, and that it will hence flourish there in its greatest perfection."

It will be seen that, though it is claimed that the cane will flourish best in the South, yet it is freely allowed that it will do well wherever Indian corn will flourish. If any doubts still exist in the mind of the reader, I hope to be able to remove them, so far as possible, before I finish.

I shall next give an article from Prof. J. J. Mapes, of Newark, New Jersey, which appears in the November number of the *Working Farmer*, entitled "Refined Sugar from the Chinese Cane." Prof. Mapes is known to be *au fait* in all such matters, as we shall see from the valuable and practical matter below.

"During the past season many new facts have occurred which cannot but interest our readers. Among these is the successful cultivation of the Chinese Sugar-cane, or Sorgho. We received a small package from the Farmer's Club of the American Institute, and have grown a few square rods. Messrs. Olcott and Vail, of the Westchester Farm School, at Mount Vernon, New York, have raised an acre, and both their experiments and our own, so far as pursued, seem to endorse the views of others. The stalks of the sorgho are more numerous than those of corn, and grow with us eleven feet high. The quantity of seed is very large, while the stalks and leaves are much sweeter than corn-stalks, and are readily eaten by cattle, being preferred by them to the stalks of the sweet-corn. Messrs. Olcott and Vail have made syrup from the juice, of a light straw-color, and in every way equal in flavor to that of the sugar-cane. If our friends in Carolina are right in their views of the value of this plant,— and we have no reason to doubt them,— it will enable a large portion of the Northern States to manufacture sugar of good quality — indeed, of any quality, as, from the improved method now understood, white sugar (refined) may be made direct from the juice. The only difficulty in this manufacture consists in the fact that large manufactories alone can produce the best qualities at low cost; and not until such factories can be established in districts capable of supplying the necessary quantities of canes can the best results be obtained. In the mean time, we will give such necessary directions for the manufacture of sugar from the juice of the sorgho as may be best availed of by the small operator.

"When the grower intends to make sugar, he should pinch off the seed-heads before they are fully formed, or, indeed, as soon as they appear, thus causing the plants to give a larger yield of stronger juice. A cheap and effective mill for expressing the juice may be made of three rollers, arranged like the ordinary sugar-mill for West India use, but of small size. Two of the rollers should be on a horizontal plane, with a third roller above, and all geared to the same speed. Such a mill will separate much of the juice, and it may be used by hand or other power, as preferred. The great art of sugar-making is to get the largest quantity of crystals and the smallest of molasses or syrup, and this will depend in a great measure on the rapidity of the process. Even the quality of the molasses itself is dependent upon its rapid concentration during the early stages of manufacture. All must have observed that a *freshly* broken or cut apple, if exposed to the atmosphere, will become brown in a short time; and a similar effect is constantly going on with cane-juice, from the time it is expressed until its final concentration.

"The apparatus for clarifying, concentration, etc., so should be constructed as to insure the greatest rapidity of action. In a small way brass kettles may be used; but for larger operations, requiring new ones to be constructed, they should be of copper. The use of alkalies in clarifying has long been known, and their excessive use often injures the quality of the results. The operator should supply himself with three kettles, two large and one small. The juice, as soon as expressed, should be placed in one of the large kettles, and to which should be added — say to ten gallons — half a tea-spoonful of cream of lime, one pound of finely-ground and freshly-burned bone-black, and two ounces of bullock's blood, or the whites of two eggs, or half a pint of skim milk, — either will do. The blood or eggs, if used, should be beaten, and then well divided throughout the mass, stirring all cold, and during the early part of the heating. The process in this kettle should be conducted somewhat slowly, and if the kettle be large enough to permit all the scum to rise without overflowing it, the scum need not be removed, as it will remain on top of the fluid, becoming more and more compact. The juice should not be allowed to boil or simmer. After the clarification is perfect, the scum on top will crack open in all directions, and white, sparkling bubbles will rise through these cracks, overflowing the top of the scum, and it will turn over in masses. The scum may be taken off and the juice thrown on a blanket in an open basket, thus partially filtering the mass.

"It should then be placed in kettle No. 2, and boiled as rapidly as possible until a thermometer placed in it will indicate 220 degrees of Fahrenheit, when it should be again filtered. The first portion passing the filter should be returned, as it will not be quite clear. The whole will then be bright, and may be put into kettle No. 3, which need be but half the size of the others, and should be placed on a clear, strong fire, and so arranged that it can be readily taken from the fire at short notice. Place in this kettle a thermometer—it will commence boiling at 220 degrees, and gradually increase to 240 degrees; the instant it reaches that point it should be taken from the fire suddenly, for if permitted to rise to 241 degrees, or more, it can never be purged. Let it stand in this kettle until a slight crust commences to form on the sides and top, then scrape this down with a wooden spatula, thin at the end and edges, and stir all until evenly mixed with the more fluid portions; then pour into a conical sugar-mould, stopped at its lower end, and place the nose of this mould on a drip-pot. This sugar-mould should be of the kind known as the Bastar-mould, and it and the drip should stand in a warm place. The next day the sugar in the mould will be solid, and the plug in the bottom of the mould may be withdrawn and an incision made upward with a pegging-awl, replacing the mould on the drip-pot. The sugar or molasses will gradually drip from the nose of the mould into the pot, and the time necessary for this purging will depend upon the heat of the apartment where it is placed; usually the syrup will all run off in the natural way in a week or ten days, leaving the sugar in the mould of a light straw-color. If the operator desires to make the sugar white, he may do so by the process of claying, or liquoring. We were several years engaged in sugar-refining, making sugar, etc., and shall be fully prepared to give all the necessary particulars both for large and small manufacture, should the experience of next year prove the practicability of Northern sugar-making.

"From the experiments made in Georgia we cannot but believe that in the Middle States, at least, this new industry may possibly be profitably prosecuted. In the large way, the expense of manufacture need not be so great as named by Mr. Peters, — ten to fifteen cents per gallon. Indeed, from our experiments with the stalks for feeding purposes, we think the unripe portions of the canes, or those not in the best order for sugar-making, added to the begass, or pressed canes, and cut up in the ordinary way, would be worth as much for fodder as the cost of the sugar-making, in such localities as can supply themselves cheaply with fuel, etc.

"It will be remembered, that while the Hon. H. L. Ellsworth was Commissioner of Patents, at Washington, he was much interested in the manufacture of sugar from corn-stalks; and in consequence of the excitement at that time we raised a quantity of sugar, or sweet corn, for the purpose of sugar-making, pinching off the fruit as fast as it appeared — and thus forcing all the secretions of the plant into the stalk. The growth was very large, and the juice highly charged with sugar, its strength indicating 10½ degrees on Baume's saccharometer, being stronger than the best Louisiana cane-juice, and, of course, capable of giving more sugar per gallon; but, unfortunately, so few gallons per acre as not to give a paying result. We made refined sugar from these corn-stalks, and that year exhibited at the American Institute several loaves of corn-stalk sugar."

I think I cannot more profitably occupy the space than by continuing these reports from different parts of the country; for by them each one will be able to judge for himself whether we have got in this cane a plant adapted to our various latitudes.

Extract from a statement of Joseph C. Orth, of Illinois, from Patent Office Report of 1855.

"Profiting by the remark printed upon the paper which contained the seeds, — 'good for fodder, green or dry, and for making sugar,' — I cut off a few stalks and offered them to my horses and cattle, which ate them with apparent good relish, and seemed to ask for more. I then concluded that, as a part of its recommendations were true, I should also try the other, and manufacture sugar from the juice. Its stalk being very long and heavy, and exceedingly rich in juice, and to the taste, in its natural state, almost as sweet as molasses, no doubt remained upon my mind that it was what it was said to be. I cut six stalks, placed them successively upon a flat board, took a rolling-pin, and, as well as this simple machine enabled it to be done, expressed and saved the juice. The result was, I obtained two tumblers-full, but half was not saved. This was then boiled down, and produced one of the same tumblers half-full of good, pleasant-tasted molasses, about as thick as the common molasses obtained in the shops. But, as my object was simply to ascertain the quantity rather than the quality of saccharine matter contained, this juice was neither strained nor clarified, and therefore its taste was not equal to what it would be under more careful treatment. From all I could observe concerning this plant, I am fully convinced that 15 per cent. of good clarified sugar could be obtained from the juice. My experiment produced about 25 per cent. of molasses. This, it would seem, is evidence strong enough to warrant a more extended trial of its merits; and if it will in any way supply the place of cane-sugar, it must of necessity become a very important and valuable acquisition to the agricultural products of the Middle and *Northern* States. I am fully satisfied that it will ripen in north latitude 42°, which is about the northern limit of Illinois."

Extract of a statement of Samuel Clapham, of Suffolk County, New York:

"Early in May last I received a small parcel of the seeds of the Chinese Sugar-cane (*Sorghum Saccharatum*), which I cultivated somewhat after the method of Indian corn. The proper time for planting, however, I should say would be the same as that of early corn, as I find it quite hardy; and stalks of it cut down the end of October made fresh shoots after two rather heavy frosts, and still were good for feed. From twenty-five plants I obtained half a bushel of ripe seed.

* * * * * * *

"Although in this part of the country I look upon this plant as of great value as a forage crop, yet possibly it may be profitably cultivated for sugar, as the juice contains nearly ten per cent. of saccharine matter as clear as crystal, and, on a very small scale, beautiful clarified sugar was produced by my friend Dr. Ray."

D. Minis, of Beaver county, Pennsylvania, writes thus : "Last spring I received from the Patent Office a small parcel of the seeds of the Chinese Sugar-cane. I planted it about the 20th of May, although it might have been sown ten or fifteen days earlier; but, fearing that it might be injured by a late frost, I preferred to plant it thus late. I planted it in the centre of a twenty-acre field in two rows, with the hills about three and a half feet apart, with from two to six seeds in each hill. Where the plants were three or four to a hill they grew the most vigorously, and seemed to produce the most perfect seed. I gave them no extra culture, either in labor or manure: the plants had no protection from sunshine or storm before I secured the seed. The given weight of the crop on a given space, growing as it did with me the past season, I think would be nearly or quite equal to that of Indian corn."

D. J. Browne, Esq., of the Patent Office, Washington, D. C., in his late report, thus speaks of the Chinese Sugar-cane : "Since its introduction into this country it has proved itself well adapted to our geographical range of Indian corn. It is of easy cultivation, being similar to that of maize, or broom-corn ; and, if the seeds are planted in May in the Middle States, or still earlier at the South, two crops of fodder can be grown in a season from the same roots, irrespective of drouth : the first one in June or July, to be cut before the panicles appear, which will be green and succulent, like young Indian corn ; and the other a month or two later, when or before the seed is fully matured. The amount of fodder which it will produce to the acre, with ordinary cultivation, may be safely estimated at seven tons when green, or at least two tons per acre when thoroughly cured. The stalks when nearly mature are filled with a rich saccharine juice, which may be converted into sugar, syrup, alcohol, or beer, or may be used for dyeing wool or silk a permanent red or pink ; and the entire plant is devoured with avidity, either in a green or dry state, by horses, cattle, sheep, and swine. Considered in a utilitarian point of view, this plant perhaps has stronger claims on the American agriculturists than any other product that has been brought to this country since the introduction of cotton or wheat. Aside from other economical uses, its value for feeding to animals alone, in every section of the Union where it will thrive, cannot be surpassed by any other crop, as a greater amount of nutritious fodder cannot be obtained so cheap on a given space within so short a period of time. Without wishing to present the question in an extravagant light, it may be stated that this crop is susceptible of being cultivated within the territory of the United States to an extent equal to that of Indian corn, say, 25,000,000 acres per annum; and estimating the average yield of dry or cured fodder to the acre at two tons, the yearly amount produced would be 50,000,000 tons, which, to keep within bounds, would be worth at least $500,000,000, besides the profit derived from the animals in milk, flesh, labor, and wool."

The above article is from the gentleman who introduced this sugar-cane into this country, and from this fact, as well as his connection with the Patent Office, is competent to judge of its merits. The evidence given by him is fully sufficient to induce every farmer to try it for fodder, if for no other purpose.

A writer in the *Chicago Free Press* expresses the opinion that "in 1860 the Southern planter will have no sale for his sugar in the State of

Illinois. From present indications, there will be one hundred acres of Chinese Sugar-cane raised in Wabash county, next year, which will save the county $10,000."

I now give the results that have attended the growing of Chinese Sugar-cane in the New England States. Most of the accounts previously given were from persons at the South and West. The reason I shall not be able to give so full an account of its culture and manufacture at the North is that it has not been so extensively tried; in fact, but few persons, so far as I am able to judge, have heretofore thought it worthy of notice. Among those who have given it a trial is the editor of the *Amherst* (*N. H.*) *Cabinet*, to whom I am indebted for the following:

"We have frequently alluded to our experiment in raising the Chinese Sugar-cane, from seed received at the Patent Office. We are entirely satisfied that it can be raised with great profit in this locality, either for fodder or for the making of sugar or molasses. We have in a small way tested it for both, and think we can satisfy the most incredulous that our farmers can raise molasses and sugar to better advantage than they can either corn or potatoes. Our seed we received late, and planted after corn was generally up. When about a foot in height, it encountered a violent hail-storm, which seriously damaged its growth, so that it is doubtful whether the seed is sufficiently ripe to be reliable; and we shall accordingly secure a supply for ourselves and others, to whom we have promised it, from Washington or elsewhere. Preferring securing the seed to experiments in sugar-making, we allowed our cane to stand beyond the proper season for the latter purpose, and after gathering it stood several weeks before used. Last week, finding election over, and no firing to do, and but little to interest us in the papers, we essayed to convert the product of six hills, planted like corn, into molasses. We run thirty-two stalks through a hay-cutter, and with our standing press and a cheese-hoop took therefrom three quarts of clear and rich juice, which, being boiled to the consistence of sugar-house molasses, yielded one pint. The flavor is very agreeable, and the color and appearance nearly that of honey; and it is the universal opinion of those who have tested it that it is superior to any Southern molasses."

I will next give the success of one of my neighbors, and an esteemed friend, Mr. Jonathan Stone, of East Newton, who raised some of the sugar-cane the past season. It was quite late before he received the seed, so that it was not planted until about the first of June. It was put into hills, in a cabbage-field, where the cabbages had failed. It received but little attention until quite late in the autumn, after there had been several frosts, so that the leaves were all killed. About the middle of October, Mr. Stone, at my suggestion, expressed the juice from a number of stalks, boiled it down, without using lime or any other substance either to clarify or to neutralize the acid, and obtained a beautiful article of syrup, such as my own cane furnished. The seed did not ripen well, on account of the lateness of planting. The cane grew in this short time to over ten feet in height. It will be seen by this that the cane may be, if planted early in the season, grown even in Massachusetts, so that there would, ordinarily, be a whole month to manufacture sugar or syrup. We also learn, by this, that no extra amount of manure or labor is necessary to raise the cane of good size.

Mr. John A. Kenrick, also of Newton, raised some of the sugar-cane. He started it in a hot-bed, and then transplanted it into hills in the field, where it grew to the height of eleven feet, and very stout, the best specimens I have ever seen. It ripened its seed well.

A correspondent of the *Boston Atlas* communicates the following, concerning an experiment made in this state with this plant: "The seeds were planted on the 14th of last June, in the same manner as corn is usually planted. The canes grew to the height of ten feet, and spindled like broom-corn, but did not ripen any seed. About the 10th of October, the crop of about forty canes was harvested. The juice was expressed by means of a sugar-mill, such as used by grocers for crushing sugar. The forty canes yielded about two pints and a half, which was boiled down to syrup."

From the *New England Farmer* we clip the following, written by J. J. H. Gregory, of Marblehead, Mass.: "About the middle of last June I received a small package of the seed of the Chinese Sugar-cane. On the 18th of the same month I planted a few seed for experiment, from which one hill of seven plants was reserved. These thrived well, and at the time of the first heavy frost had attained a growth of about ten feet, with the seed at the tops apparently full-sized, but, as was anticipated from the shortness of the growing season, not well filled, and scarcely colored. From six of the stalks the juice was expressed and boiled down to the consistency of common molasses, yielding about a common coffee-cup full (or one and two thirds gills) of a rich syrup, which our grocers considered to be richer flavored than ordinary molasses, equal in quality to the syrup of commerce. Please accept, with my best wishes, Mr. Editor, the accompanying sample of the molasses."

The editor says: "Friend Gregory will please accept thanks for his fine specimen of Chinese sugar-cane molasses,— an article, we trust, yet to be generally introduced as one of our staple New England crops. It is a syrup rather than molasses, the latter being an article *drained* from sugar. Let it have a more extensive trial, another season."

A gentleman in Dorchester, Mass., grew the cane last year, and the seed ripened some weeks before the frost came. This goes to show what I have said in another place, that in a good season, when we have warm weather,— warmer than the past season has been,— this plant would fully mature by the 1st to the 10th of September. It is true this would give but a short time to manufacture a quantity of sugar or syrup, unless a great many hands were employed. And this would involve a large outlay for a mill, or mills, kettles, &c., which would be indispensable on a large place. The climate of Louisiana allows the planter from sixty to ninety days to secure his crop; and even this often proves too short, for the frosts come on, and put a stop to the sugar operations.

I find the following in the *Massachusetts Ploughman*, in regard to the new cane, and the syrup manufactured therefrom. It is from Mr. Foster Bryant, of Mansfield, Mass., a gentleman well known to the public:

"To the Editor of the Ploughman. — Sir: With this I send you a sample of syrup obtained from the Sorgho Sucré. I received a package of seeds from the Patent Office, which I planted in hills three feet apart in a single row, and on land varying from the capacity of 14 hills to a bushel of potatoes up to 60. On the good land the plants attained the height of ten feet, while on the poorest the height did not exceed three feet six inches. I suffered six stalks in a hill to grow. Hoed but twice; planted in the middle of May; land moderately manured broadcast, but not in the hill. I could not obtain rollers to crush the stalks, and therefore resorted to the expedient of splitting and boiling in water.

I do not suppose I obtained more than half of the saccharine sap by this method; while the added water greatly protracted evaporation, which, having been performed in an iron kettle, has probably heightened the color, and for aught I know imparted a somewhat peculiar taste. I can give no information touching the cost of the syrup; and, being ignorant of the art of making sugar, I made no attempt to procure the latter.

"I think land that will grow fifty bushels of corn to the acre will bear a heavy crop of the sorgho sucré; but I very much doubt if our season is long enough to ripen the seed, unless the plants are brought forward in a hot-house. In my case, the seeds were not fully formed when the heavy frosts occurred, the first of which made *ice* one quarter of an inch thick, *without white frost*. This apparently did not injure the plants. The following night a heavy white frost occurred, and the leaves were shrivelled and dry before midday, but the stalks showed no change. I cut them the same afternoon."

The editor comments as follows: "We have received a bottle of the saccharine matter from Mr. Bryant, and given it a taste. It is of much the same flavor as the sap of the sugar-maple, when boiled down to the degree of consistence.

We thank Mr. B. for his specimen of the sap thus prepared ; and hope others will be induced to grow this plant, on a small scale at first, for it may prove a superior article for fattening animals, if not for making sugar.''

Notwithstanding Mr. Bryant did not succeed in ripening the seed, many others have, even in Massachusetts, so that we need not doubt on that score. And even if the seed should not all ripen, or any part of it, the cost of seed would be but little ; for it can be imported in great abundance from France, or brought from the South, so that it will undoubtedly be furnished in a year or two for less than fifty cents per pound. It will also be seen by the above that the stalks did not receive any apparent injury, even from a heavy frost, though I very much doubt whether it would be best to run the risk. It has been found to flourish as far north as Minnesota, where it has attained the height of twelve to fifteen feet. So in the New England States, where it has not only grown well, but ripened its seeds. It has been grown to some extent by many persons through the New England States, as well as in other portions of the country, and the universal testimony is, "It grew well with us, but we did not try it for sugar or syrup." I have not thought it best to take more space in giving accounts of experiments, presuming that enough have already been given.

The following brief hints may be of use to those who propose to enter upon its cultivation :

1. Select a warm and dry soil, such as you would select for Indian corn.

2. Prepare your ground precisely as you would for corn, either by spreading your manure, or putting in hills,— about the same distance between the hills, where the ground is rich.

3. In planting, which should be done early, put into each hill six or eight seeds. Cover lightly with well-pulverized soil,— say, three fourths to one inch deep ; pull out all but four or five at second hoeing. If planted in drills, seed enough should be used so that after hoeing there may be a stalk to every four or five inches ; from a pound and a half to two pounds of seed should be used.

4. Cultivate and hoe as with corn ; care should be taken that the ignorant do not hoe up the young plants, taking them for barn-grass, which they very much resemble

5. When the panicles appear, they should be cut off of all that which is intended for sugar or syrup making.

6. When the plant has just passed into bloom, the stalk may be used for syrup, but will continue to grow better until the seed is in the milk-stage, or little later.

7. The stalks should be cut close to the ground, with a bill-hook or some such tool, and stripped of their leaves, and the green, succulent top cut off, when they are ready for the mill; the leaves and top may be fed green to cattle, or dried.

8. The stalks should be passed through the mill twice or more, until most or all of the juice is expressed.

9. The juice should not be allowed to stand long after being expressed, but boiled at once, if possible. A slow fire should be made under the kettle, — which should be of brass, or much better of copper, — and the juice should not be allowed to boil until the green scum has all been taken off. Lime-water may be used to aid in clarifying and to neutralize the acid; the exact quantity is not yet determined, but to every five gallons of juice, say, from one to two tea-spoonsful of powdered lime, or the same dissolved in

water, and strained, before being put into the juice.

10. When all the green scum has been removed, the fire may be increased, and the juice boiled down until nearly as thick as common molasses in hot weather, when, if intended for syrup, it should be removed from the fire, for this completes the process. If intended for sugar, it should be allowed to boil longer, and until it will "string into threads," or present an appearance of being sufficiently boiled to grain, when it should be thrown off into troughs, or coolers, at once. I am not able to give exact information in regard to the time it should be boiled to crystallize readily. Further experiments will determine.

11. If made into sugar, it should be removed from the coolers to casks with holes bored in them, so that the molasses may drain off and leave the sugar dry, as it should be. These casks are generally placed on timbers, with a cement cistern underneath to hold the drippings, or molasses. After remaining in the "purgery" until sufficiently drained, it comes out fit for sale, or use.

12. If cultivated exclusively for fodder, it should be planted as early as the weather will

allow, and quite as thick as stover-corn. When the panicles appear, or even before, it may be cut either for soiling or for drying, and the roots will at once throw up another crop.

13. If it is desired, the juice may be fermented, like the juice of apples, being put into casks at the mill, and treated like cider.

14. The begass, or waste, may be dried and used for fuel, or for making paper, or rotted down for manure.

15. If the storms should blow down the seed-cane, no fears need be entertained, as it will remain weeks in that condition without injury. I must here caution all persons who grow this cane against planting it in the vicinity of broom-corn, Dourah-corn, or Guinea-corn; for it readily mixes with these plants, and it would render the seed worthless for planting.

I think I have sufficiently shown that the Chinese Sugar-cane may be grown, both North and South, with success, either for sugar and syrup making, or fodder, or some of the many other uses to which this wonderful plant is adapted. It may be, and doubtless is true, that the climate of the South is better adapted to the production of sugar, inasmuch as there will be a greater amount of crystallizable sugar obtained from the same amount of juice than at the North; and also that the seed will be more sure of ripening, and better in some respects after it is ripened, may also be true. But this should not prevent the North from engaging in its culture, by any means; for, even though it should cost as much to produce sugar on our own farms as we could buy it for in the market, yet I believe very many would choose to make their own; for, aside from the fact that many would prefer sugar produced on free soil by free labor, they would rather make it themselves, and thus turn their labor into money, than pay out the ready cash for an article grown in a foreign country, or even at the South. But, should it turn out, after it has been fully tried, that we cannot profitably make sugar from this cane at the North, then I take it the cane would be grown for syrup-making; and the only possible thing there will be to prevent this — for I think I have shown beyond contradiction that this can be done at the North — may be that the South can produce as good syrup, and deliver it at our doors for a much less price than we can do it. But I don't believe that that can be done; for I see no reason why we should not be able to compete with them in this article, for certainly they will labor under the disadvantage of bringing it here for a market, while we shall have ours on the very spot.

After the mill has been set up and the boilers arranged, there will be but little expense except for labor and fuel. In regard to the former, I believe free labor can and always will compete successfully with slave labor, give it an equal chance ; in regard to the latter, the begass may, and possibly will, be used for fuel in some places, as in the West Indies, where it supplies nearly all the fuel required both to run the steam-engine and to boil the syrup. I believe the time will come when we may revel in sweets grown upon our own free soil, either from this cane or other saccharine plants that will be introduced. Glorious results are to follow the introduction of this plant, if all our anticipations are realized, when the poor as well as the rich shall have the sweets of life within their reach ; for it is the *masses* we would benefit. The rich can obtain sugar, let it cost what it will; but not so with the laboring man,—he must be deprived of this luxury, if the prices advance as they have for the past two years. But let us not get excited on this subject, so that the Chinese Sugar-cane excitement will be classed with the Merino sheep fever, the Morus multicaulis,

Rohan potato, Fowl, and other fevers that have had their day, and are only recollected as specimens of our folly as a people. This new plant is no humbug, but I believe a downright valuable article. Careful experiments yet to be made will determine *how* valuable. Let each and every agriculturist try it another season,—on a small scale, if he chooses, but at any rate try it and judge for himself. I shall plant at least an acre, and with a perfect sugar-mill, and other apparatus which I mean to obtain, I shall make thorough and careful experiments, which will be given to the public at the end of the season. Many there are who stand ready to denounce this, as all new things, as a humbug, and a worthless article. To such I say, suspend your judgment until a fair trial has been made, until it has been proved worthless, and then I will join with you heartily in denouncing it.

In closing, let me say that my object has been to give the reader all the facts within my reach; and, though I may not have succeeded in giving all the information that may be needed, the reason, I think, will be apparent to all. It being a new thing, of which little was known until its introduction into the United States, and the time has been so short for us to experiment with it that it is not in the power of any one to give accurate rules for its cultivation, treatment, &c., or to speak positively of its properties and merits. I have given briefly some rules in regard to its culture, &c., but they must necessarily be imperfect, as they are founded only on my own experience of one year, with what I have obtained from others who have grown it during the past season. I have not intended to over-color the remarks I have made; and in the selections I have given both sides are shown, — the dark as well as the bright. I have been able to obtain letters from distinguished gentlemen, fully competent to speak on this subject, and who have had some experience. Their opinions, you will agree with me, are entitled to great weight in this matter.

LETTER OF HON. MARSHALL P. WILDER, PRESIDENT OF THE UNITED STATES AGRICULTURAL SOCIETY.

Dorchester, Dec. 19th, 1856.

J. F. C. HYDE, ESQ.

DEAR SIR: Your favor, requesting a word from me in relation to the New Chinese Cane (*Holcus Saccharatus*), the product of which was exhibited at the late show of the United States Agricultural Society, is received. Several samples of syrup made from this cane were presented by Col. Peters, of Georgia; and, in my judgment, it was one of the most important articles on exhibition, connected with agriculture. No subject has excited more deep and universal interest throughout the country, for many years, than the introduction of this new plant. Col. Peters is therefore entitled to great credit for the exposition of this article in this public manner, and also for the detailed statements which he gave of his method of cultivation, manufacture, and the result of his operations. I was happy to learn, on my return home, that yourself and others had been equally successful with that gentleman, and that it is your intention to give the subject further investigation. This cane has been grown during the past season with as much success in the Eastern and Western States as in Georgia and the extreme South, and presents to our farmers the prospect of producing their sugar and molasses as easily as almost any other crop. It is capable of being cultivated wherever Indian corn will succeed, and, of course, to the same extent, and to a much greater profit. Col. Peters writes me that he shall plant one hundred acres next year; others are proposing to plant largely. If it can be manufactured into sugar, or molasses, of which there seems to be no reasonable doubt, it is impossible to predict the importance of this crop to American agriculturists, or to the country at large. Go on, my dear sir! You are on the right track. This is not the only species of the *imphees*, or sugar-cane, to be brought to notice. There are other varieties in Caffraria and Algeria, which are said to be very superior; and it is to be hoped that the vessel which has recently been sent out by the government of the United States will not return without bringing a supply of these plants, or the seed of them, so as to place at an early day before our yeomanry all the information that can be obtained on this most interesting subject.

Yours, with great respect,
MARSHALL P. WILDER.

The following letter from Gov. Gardner will be read with interest.

Boston, 18*th Dec.*, 1856.

JAMES F. C. HYDE, ESQ.

MY DEAR SIR: I hasten to reply to your note of the 16th, relative to my experience and impressions regarding the susceptibility of the cultivation of the Chinese Sugar-cane in our climate.

In the autumn of 1855, I learned that an esteemed friend, and a neighbor in the summer season, Benjamin Hemmenway, Esq., of Dorchester, had grown some sugar-cane upon his lands, and that it had matured and given evidence of being well stocked with saccharine matter. Feeling an interest in the subject, I applied for some seeds of his own growth, which he kindly gave me.

I planted them in hills, *quite late in June*, 1856. I confess they were put into the ground so late in the season I did not expect them to reach maturity, and my chief object was to know if seeds grown *in our latitude* would ripen sufficiently to germinate and produce full-sized cane. There was no doubt that exotic seeds, brought from warmer climates, would grow more or less perfectly the first year; but it is a totally different question whether the seeds of such plants will again sprout and grow to perfection.

Not anticipating my cane would ripen, I took but little trouble in planting the seed; and it is worthy of note, that it was planted in a tolerably rich loam, but *without any manure*. In a short time the plants appeared, looking like hills of corn, and nothing was done for it excepting keeping down the weeds, saving that it was moistened three or four times with a weak solution of guano and water. I planted thirty hills in two rows, five seeds in a hill, and about the same space between the hills that is adopted in planting corn, one end of the rows running under a very large elm-tree. The cane grew with great rapidity, but there was soon a very obvious difference between those hills that were shaded and those that were wholly exposed to the sun; and, in the end, not only a variation in the height of nearly one half, but those that were the most shaded failed altogether to spindle out into seed-blossoms.

The extreme height of the most favored hills must have ranged from *ten to twelve feet*, judging by the eye, and nearly or quite every seed planted threw up a stalk. Notwithstanding I had not hoped, owing to the period of its planting, it would mature, much of it did; and, though an early October frost checked it (for it was on low land), a good part of the seed ripened, and I propose planting some of it the next year.

I think the following facts are satisfactorily developed from my experiments:

First. That seed grown here will produce plants as perfect as the imported seed: in other words, that the cane can be perfectly *acclimated in our state.*

Second. That it will probably mature in any season when Indian corn will.

Third. That it requires a sunny exposure, as corn does.

Fourth. That it does not need excessive artificial fertilization: or, in other words, does not excessively exhaust and impoverish the soil.

Fifth. That, as a green fodder, it produces more food for cattle on same space and at same cost than corn.

Sixth. That cattle prefer it to corn fodder; for I repeatedly gave it mixed with corn-stalks to cows, and it was amusing to see them carefully select the sugar-cane from the corn-stalks, eating the former first, as I have seen hogs pick out pears from apples.

Seventh. That it is much more juicy and nutritious for milch cows than any other fodder; for it is well known that sugar contains more nourishment than almost any vegetable production in daily use.

You will notice I have made no reference to the possibility of this interesting plant containing sufficient saccharine matter to enable it to be converted into sugar and molasses at such rates as to successfully compete with the sugars of Louisiana, Texas, and the West Indies. Yet there is no doubt but that it far surpasses the sugar-beet in this respect, which has for many years in France produced those articles profitably; and as little do I question but that it possesses more properties of the Caribbean cane-syrup than the maple, from which considerable quantities of sugar are annually manufactured in various parts of New England.

In fact, it may be found that it is as well adapted to the manufacture of these necessary articles of domestic economy as the cane of our Southern States. Should such prove to be the case, an immense industrial revolution is at our doors, the results of which must be as gratifying as stupendous. Many millions of dollars, doubtless, are annually sent away from New England to purchase Southern sugars, which will then be kept at home to enrich the producer upon the hill-sides and in the valleys of our section of country. And, better than all, one great staple, which is almost the exclusive growth of slave labor, which props up that institution and adds to its continuance, will be wrested from its tottering basis.

Vigorously pursue any practical course of economic effort which will tend to make slave labor less profitable, and you do more to bring about that prophetic and certain day "when bondage shall exist no longer," and "the enslaved shall go free," than by all the refinement of political ethics, or even the crushing influence of exotic humanity, Christian sympathy, and popular sentiment.

In the latter view of this question, especially, is it our duty to pursue a thorough and systematic course of experiments, to fully ascertain the capabilities of this new plant. I rejoice that a gentleman of your perseverance and intelligence is determined to aid in accomplishing this end. Count me not only as your well-wisher, but pecuniary aider, if necessary; and I earnestly hope success may crown your efforts.

I omitted to say the specimen of molasses of your *own growth and manufacture* from this cane duly reached me. It tastes deliciously, and looks promising, realizing the proverb that "the product of one's own labor is *sweet*."

Should any facts in this note be of service to you, please make what use you please of them, and believe me

Very truly yours,

HENRY J. GARDNER.

FROM CHARLES L. FLINT, ESQ., SECRETARY OF THE MASSACHUSETTS BOARD OF AGRICULTURE.

State-House, Boston, Dec. 22d, 1856.

DEAR SIR:

I have had opportunities of various kinds to learn something of the success which has attended the culture of the Chinese Sugar-cane, and am happy to hear that you are preparing a treatise upon it which will embrace, no doubt, all that is at present known among us of its natural history, its comparative value, and the best modes of cultivating it. Such a work, I am sure, is greatly needed as a guide for our experiments in the introduction of a plant new, at least, to us, and which promises to prove so valuable. Many a new plant and many a new implement of husbandry is thrown aside after a feeble effort, when a little knowledge of its uses and value on the part of the experimenter would have led to an entirely different, and, perhaps, perfectly successful result. The manual which you contemplate will, therefore, come just in season.

To say of this plant that it will work an entire revolution in the great sugar interest of this country, would, perhaps, be premature; but the fact that it has sprung into general notice and awakened the interest of the whole country so suddenly, is strong evidence of its intrinsic importance, especially as it has succeeded in every case, so far as I know, beyond the anticipations of those who have tried it. It bids fair to become of national importance.

I have some acquaintance with Col. Peters, of Georgia, whose statements are before the country. These statements are perfectly reliable, and show what we may expect from the plant in a southern latitude. Fine molasses has been made from it in Minnesota; while several experiments which have fallen under my notice in this state have been attended with success. It has been known and cultivated in France for some years.

It appears to grow luxuriantly in all latitudes suitable for Indian corn. It is not claimed, I believe, that the percentage of saccharine matter is so great in northern as in southern latitudes. This may affect its value for the production of sugar in our climate, but does not essentially affect its value as a farm product, — and especially as a forage plant, since it is, without doubt, very rich in saccharine and nutritive matters in the highest latitudes at which it can be grown. I am told, by those who have raised it, that cattle are so fond of it that they will even pick it out stalk by stalk when mixed up in a bundle of Indian-corn stalks. There seems every reason to believe, therefore, that as a forage plant it will very rapidly come into general favor, and help us essentially through our summer droughts.

Of its value for syrup or molasses I need not speak. You have shown, I believe, that it is practicable and profitable to grow it for that purpose alone. But whether it is or not, it is at least worthy of extended and careful experiments, which, I am sure, will be made, and experiments will soon determine the rank to which it is entitled among our New England productions. Your manual will encourage and assist these experiments, and it will be of essential service to the community.

Very cordially and truly, your obedient servant,

CHARLES L. FLINT.

To J. F. C. HYDE, ESQ.

FROM HON. SIMON BROWN, EDITOR OF THE N. E. FARMER.

Office New England Farmer, Boston, Dec. 27, 1856.

DEAR SIR: I am glad to learn, through your note, that you are preparing, and about to publish, a manual relating to the cultivation of the *Chinese Sugar-cane*, and the best mode at present known of obtaining the juice and converting it into syrup.

* * * * * * *

The introduction of the sugar-cane is only another of those blessings conferred on the progress of the race which have been so frequent and beneficial, and which there is still reason to believe will be greatly extended. More earnest, intelligent, and scientific investigation into the great art of agriculture will undoubtedly introduce new vegetables, and grasses, and grains, of permanent value, and new and delicious fruits, of which we are now entirely ignorant. Nature is prolific and bountiful throughout her wide realm; her secrets are not all past finding out. Intelligence and application will reveal them, and constantly confer new comforts upon all.

I hope our people will find in your manual encouragement to make multiplied experiments in the cultivation of the cane, and the production of syrup, so that out of all the trials instituted a sufficient number of reliable facts will be obtained to settle the question whether it can be produced on the farms of New England and the West at a cheaper rate than we can obtain sugar and molasses by raising other crops and exchanging them for these articles.

My own experiments in growing the cane have been quite limited, only going so far as to sow the seed and raise the plants which perfected their seeds before frosts came. I sowed the seed about the middle of May, 1856, and the plants from it perfected their seeds the first week in September. I made no attempt to express the juice, and the plants were fed to my stock when I was away from home.

I have seen numerous accounts of the growth of the cane in different parts of the country, which are all favorable to its cultivation. Bottles of syrup have been sent to me from places widely remote from each other; and those who have obtained it express the opinion that the introduction of the plant will eventually enable us to supply the market to some extent with the important staples of molasses and sugar. Their conclusions seem to me to be well grounded.

If the cane does flourish here, upon trial, our ingenious mechanics will soon manufacture mills of various descriptions, to meet every want of the cultivator, and at a cost within the means of every neighborhood, at least; so that there is every encouragement to make the experiments, in which your manual will be an important guide.

I am, very truly, yours,

SIMON BROWN.

J. F. C. HYDE, ESQ.

www.ingramcontent.com/pod-product-compliance
Lightning Source LLC
LaVergne TN
LVHW021429110826
845150LV00007B/2153

* 9 7 8 1 4 2 5 5 0 7 0 3 9 *